I0032692

Joseph William Howe

**Emergencies and How to Treat Them**

The Etiology, Pathology and Treatment of the Accidents, Diseases and Cases of

Poisoning which Demand Prompt Action

Joseph William Howe

**Emergencies and How to Treat Them**
*The Etiology, Pathology and Treatment of the Accidents, Diseases and Cases of Poisoning which Demand Prompt Action*

ISBN/EAN: 9783337778934

Printed in Europe, USA, Canada, Australia, Japan

Cover: Foto ©berggeist007 / pixelio.de

More available books at **www.hansebooks.com**

# EMERGENCIES.

### THE ETIOLOGY, PATHOLOGY, AND TREATMENT OF THE ACCIDENTS, DISEASES, AND CASES OF POISONING WHICH DEMAND PROMPT ACTION.

DESIGNED FOR

## STUDENTS AND PRACTITIONERS OF MEDICINE.

BY

### JOSEPH W. HOWE, M. D.,

AUTHOR OF "THE BREATH, AND THE DISEASES WHICH GIVE IT A FETID ODOR;" "WINTER HOMES FOR INVALIDS," ETC.; LATE PROFESSOR OF CLINICAL SURGERY IN BELLEVUE HOSPITAL MEDICAL COLLEGE; FELLOW OF THE NEW YORK ACADEMY OF MEDICINE; MEMBER OF THE NEW YORK SURGICAL, PATHOLOGICAL, AND COUNTY MEDICAL SOCIETIES; CONSULTING PHYSICIAN TO HOSPITAL FOR DISEASES OF THE NOSE, MOUTH, AND THROAT; VISITING SURGEON TO CHARITY AND ST. FRANCIS HOSPITALS.

*FOURTH EDITION.*

NEW YORK:
D. APPLETON AND COMPANY,
1, 3, AND 5 BOND STREET.
1890.

# PREFACE TO THE FIRST EDITION.

THIS volume, as its title indicates, is designed as a guide in the treatment of cases of emergency occurring in medical, surgical, or obstetrical practice. I have endeavored to combine, in a narrow compass, all the important subjects, giving special prominence to points of practical import in preference to theoretical considerations, and, with the results of my own personal observation, uniting the latest views of European and American authorities.

J. W. H.

36 WEST 24TH STREET, *June* 1, 1871.

# CONTENTS.

## CHAPTER VI.

### *POISONED WOUNDS.*

## CHAPTER VII.

### *EXTRACTION OF FOREIGN BODIES.*

## CHAPTER VIII.

### *BURNS AND SCALDS.—EFFECTS OF COLD.*

## CHAPTER IX.

### *STRANGULATED HERNIA.*

## CHAPTER X.

### *LOSS OF CONSCIOUSNESS.*

#### COMA.

## CHAPTER XI.

### *LOSS OF CONSCIOUSNESS*—(CONTINUED).

#### SYNCOPE.

## CHAPTER XII.

### *ASPHYXIA.*

## CHAPTER XIII.

### *SUNSTROKE.*

## CHAPTER XIV.

### *DYSPNŒA.*

## CHAPTER XV.

### *ŒDEMA GLOTTIDIS.*

## CHAPTER XVI.

### *CONVULSIONS.*

## CHAPTER XVII.

### *SUSPENDED FŒTAL ANIMATION.*

## CHAPTER XVIII.

### *COMPLICATIONS OF LABOR, ETC.*

## CHAPTER XIX.

## CHAPTER XX.

### TOXICOLOGY

#### NARCOTIC POISONS.

## CHAPTER XXI.

### IRRITANT POISONS.

## CHAPTER XXII.

### METALLIC POISONS.

## CHAPTER XXIII.

### CORROSIVE ACIDS.

## CHAPTER XXIV.

### CORROSIVE ALKALIES.

# EMERGENCIES,

AND

# HOW TO TREAT THEM.

---

## CHAPTER I.

*HÆMORRHAGE.*

General Considerations.—Results of Negligence.—Arterial and Venous Hæmorrhage.—Effects of Profuse Hæmorrhage.—Natural and Artificial Methods of suppressing Hæmorrhage.—Hæmorrhagic Diathesis.—Transfusion.

MEDICINE is often reproachfully characterized as a science of experiments, a profession remarkable for its brilliant uncertainties and conflicting theories. Superficial observation and imperfect means of study gave origin to this sentiment when the healing art was in its infancy, and it is yet retained by a few who find it a convenient excuse for all their errors. There are conflicting theories in medicine as well as in other professions. Such theories are the mainsprings of progress; they develop strength and incite to laborious investigations.

Uncertainty appertains to every science that has not arrived at its maximum development: but it is not especially characteristic of our profession. The discoveries of to-day will necessarily be modified by the developments of to-morrow, and the theories of our own time will be replaced by the truths of the future.

The cases of emergency, considered in the following pages, are entirely exempt from the charge of uncertainty; but they are followed by disastrous results when treated by incompetent persons. The *internes* of our large hospitals know that it is not an uncommon occurrence for patients to be admitted in *articulo mortis;* their chances of recovery destroyed by the neglect or ignorance of the attendant outside. The following cases from my hospital note-book may be of interest as examples:

CASE I.—Martin C., aged twenty; occupation, machinist; was admitted to ward eleven, Bellevue Hospital, suffering from incised wounds of the wrist and palm of the hand. On arriving at the hospital, he was partially insensible from loss of blood. The voice could not be raised above a whisper, and the face was extremely pallid and anxious. The pulse could with difficulty be distinguished. The entire clothing of one side of the body was saturated with blood. On making an examination, I found that a folded handkerchief was bandaged over the centre of the wrist, and that the wound in the palm was untouched. The pad was placed on the wrist as if the greatest care had been exercised to avoid pressing on the radial or ulnar arteries. The sides of the pad scarcely reached them. This dressing was applied by a so-called surgeon shortly after the wounds were inflicted. The hæmorrhage in this case was easily controlled by ligatures. The patient, however, developed phlegmonous erysipelas, and, not having sufficient vitality to carry him through, succumbed on the fifth day after his admission.

CASE II.—John M., aged thirty; occupation, laborer; fell

from the front platform of a car at Harlem and had his right foot crushed by the wheel. His friends carried him to a surgeon in the neighborhood, who placed an ordinary bandage on the limb, without any compress over the vessels. In bringing the man to the hospital, the rough jolting of the carriage set the wound bleeding, and by the time he reached his destination he was apparently lifeless. The vessels were tied, and stimulants administered, but he never rallied. Death occurred six hours after his admission. His injuries, independent of the hæmorrhage, might, indeed, have terminated his life; still the chances would have been in his favor if a compress had been applied to the limb to prevent bleeding. The fact that such a thing was not done showed either culpable negligence or deplorable ignorance. It is through such treatment that the percentage of deaths from accidents is increased to an unnecessary degree. To remedy these evils, a thorough knowledge of the treatment of accidents should be impressed on the memory as indelibly as are the letters of the alphabet. Nor should this knowledge be entirely confined to medical colleges and professional men. Non-professionals, with a moderate share of common-sense, might learn to control hæmorrhage, relieve syncope, extract foreign bodies, resuscitate the drowned, and administer an antidote in cases of poisoning. Such knowledge would assist, rather than retard, the labor and usefulness of professional persons.

The varieties of hæmorrhage constitute a large and important class of emergencies. Loss of blood, when profuse, is always attended with danger, and necessitates immediate treatment.

The term "hæmorrhage" is applied to a flow of blood

from any part of the vascular system, with or without rupture of the vessels.

Arterial hæmorrhage is attended with serious consequences. It is readily recognized. The blood is of a bright-
scarlet color, and is forced out in successive jets; each jet is
synchronous with the movements of the heart. This characteristic spurting is caused by the intermittent force-pump
action of the heart driving out the blood. Venous hæmorrhage is distinguished from arterial by the dark-blue color of
the blood, which never flows in repeated jets, but oozes
slowly from the wounded surface. Venous blood is travelling toward the heart, and there is consequently no force behind to cause a more rapid flow. This form of hæmorrhage
is comparatively harmless, unless occurring from very large
veins.

In large wounds, arterial twigs are divided, and arterial
bleeding predominates. In small wounds there is mixture
of both varieties. The blood is dark red, and comes away
gradually.

The constitutional symptoms accompanying external or
internal hæmorrhage are distinctly marked. The lips and
cheeks rapidly assume a pallid hue. There are great restlessness and anxiety. The extremities are cold, and often bathed
in clammy perspiration; respiration is weak and sighing;
the pulse becomes small and rapid; its increased rapidity
being due to the efforts of the heart to make up, by frequent
impulses, the diminished quantity of blood sent to the tissues. The patient complains of vertigo and dimness of
vision, is unable to articulate plainly, and finally lapses into
a state of unconsciousness. The heart has partially suspended its movements, and the pulse is imperceptible. With the

syncope the bleeding ceases. There is not sufficient vitality remaining to force more blood from the injured vessels, nor action in the heart to keep up the circulation. Here Nature takes the place of surgical skill. The stoppage of the current allows the blood time to coagulate in the mouths of the bleeding vessels, and to plug them up completely before consciousness is restored or the heart again at work. But, should this fail to occur, the signs previously enumerated are intensified. A slight convulsive movement ensues, and the patient dies. Occasionally, death occurs during a sudden effort of the patient to sit up in bed, or in some other active movement. The effort creates a necessity for increased action of the heart, which is unable to respond to the call, and paralysis of the organ results. The same thing takes place sometimes when persons are greatly debilitated by disease; in rising to dress, or crossing the room quickly, they drop dead. The pulsations are abnormally multiplied, as in the former case.

There is a peculiar condition of the system known as the hæmorrhagic diathesis, in which the slightest scratch or wound of any description produces persistent bleeding. The disease is hereditary, and both sexes are equally liable to it. In Germany, beyond other countries, the largest number of cases have appeared. Seemingly insignificant wounds in persons of this diathesis endanger life. Lacerated wounds of the gums from extraction of teeth or abrasions in mucous canals, which cannot be reached by local applications, are the most serious. The blood does not exhibit the usual tendency to coagulate. The cut vessels are lax and patulous, their contractile power is diminished, and the principal natural means of suppressing hæmorrhage are unavailable.

Our knowledge of its pathology is limited, and chemical analysis shows that the blood possesses the same elements, in normal proportions, as it does in persons entirely free from this disease. The vascular canals in one or two instances have been found thinned, but in the majority of cases there is no marked alteration.

The general treatment of hæmorrhage, when thoroughly understood, can be applied in special cases without difficulty. In this connection it will be well to consider Nature's methods of closing bleeding vessels, before we pass to the appliances of art. Our efforts copy Nature as far as possible:

1. There is contraction of the muscular fibres in the artery, induced by the injury and by admission of air. The contraction closes the wounded orifice.

2. The artery retracts within its sheath, the effused blood coagulates in front of it, and the hæmorrhage consequently ceases.

3. The blood may collect on the surface, coagulate, and compress the wounded vessel.

4. If the cut vessels are small, the bleeding will cease by coagulation of blood within them.

5. Syncope, by allowing coagulation to take place before the circulation is renewed, prevents a recurrence of the bleeding.

In all our surgical methods of stopping external hæmorrhage, there are none more efficient or available than *pressure*. It can be employed over the main artery of the limb, between the wound and the heart, or directly upon the wounded part. When the main artery is to be compressed, an instrument called the *tourniquet* is generally used. If this is not at hand, a *field tourniquet* may be applied in the following manner: A handkerchief is passed loosely around the limb above the wound, and its ends fastened together. A small block of wood, a folded towel, or any substance from which a firm pad can be extemporized, is placed over the artery and under the handkerchief encircling the limb. A stick measuring five or six inches in length is then passed under the handkerchief at right angles, and twisted around until the pad compresses the artery firmly. Turning the stick draws the handkerchief very tightly around the limb and over the artery, so that it is thoroughly secured.

Bleeding from the upper extremity, at any point below the axilla, may be temporarily suppressed by placing a piece of wood an inch and a half or two inches thick under the arm at right angles with the body, and then pressing the arm firmly against the chest-walls. A large book will answer the same purpose. In all cases the material employed must be placed as high as possible in the axilla. When the wound is situated below the knee-joint, the bleeding may be diminished by raising the limb and placing it on the back of a chair, so that pressure will be made in the popliteal space. The weight of the limb in this position is sufficient to close the popliteal artery. In some cases it may be necessary to fold a towel and place it behind the knee between the chair and the limb.

Pressure may be made in a wound with the thumb and fingers, picked lint, compressed sponge, or towels. In hæmorrhage from the carotid artery, pressure may be made with the fingers along the inner edge and lower half of the sterno-mastoid muscle. The subclavian artery is compressed as it passes over the first rib, by pushing firmly with the thumb in the subclavian triangle behind the sterno-mastoid.

Pressure may be exerted on the brachial artery at the inner border of the coraco-brachialis and biceps muscles. The femoral artery is readily controlled as it passes under Poupart's ligament, midway between the anterior superior spinous process of the ileum and the pubes. The abdominal aorta may be compressed with the hand a short distance above and to the left of the umbilicus.

In wounds of the palm of the hand, or other places where there are many inosculating vessels injured, it will be expedient to place a pad or compress in the opening. Whenever the bleeding is profuse, and the main artery cannot be controlled, it is absolutely necessary to stuff the wound quickly with picked lint or other available substance. It must be filled up, packed tightly, and a bandage firmly applied. In the course of a few hours coagula may form in the vessels, when the lint may be removed and the wound properly dressed.

*Cold* is a useful adjunct in suppressing hæmorrhage. It is employed under various forms. For moderate bleeding, cloths wrung out of ice-water and placed over the part will answer. Ice in solid lumps, or pounded and secured in rubber bags, or without intervening material, is excellent in profuse hæmorrhage. Cold produced by the evaporation of ether, directed to the surface in the form of spray, has lately come into use. Prof. William H. Thompson, of this city, employs it with good results in *post-partum* hæmorrhage.

Cold acts by stimulating the arterial walls to contract, and by assisting in the formation of coagula. Cold and pressure can be used together.

*Styptics.*—Under this head are included all medicinal agents which control hæmorrhage. The most efficient are

certain preparations of iron, as the solution of the persulphate and the sub-sulphate, commonly known as "Mon-
sel's solution."

Nearly all the vegetable astringents belong to this class.
The best are tannic or gallic acids, oak-bark, catechu, and
nut-galls. Preparations of alum, and common salt (chloride of
sodium) are sometimes used. The iron and other substances
are applied by means of a camel's-hair pencil or a sponge.
They are dipped in the solution and rubbed thoroughly into
the wound. These agents act by coagulating the blood as
it flows from the vessels, and by contracting the muscular
fibres around and in their walls. The coagula thus formed
should not be removed until all danger of a recurring hæm
orrhage has passed.

*Torsion* can be employed with advantage where the ar
teries are small. This method consists in seizing the bleed
ing vessel with forceps and twisting it around until a piece
is torn off, or its orifice completely closed. Some advocate
merely one or two turns of the forceps; others believe in
tearing off a portion of the artery.

*Ligation* must be resorted to when pressure, styptics, and
milder measures fail. The ligatures in general use are silk.
The bleeding vessel is first seized with an artery-forceps, the
ligature passed around and tied. When arteries of medium
size are ligated for aneurisms, or in wounds, the middle and
internal coats are cut through, and the external wall brought
in contact. A clot forms on each side of the ligature, per-
manently closing the canal as far as the next collateral
branch. When deep-seated vessels are wounded, it is always necessary to go down in the wound and tie above and
below the opening in the artery, as the lower is frequently

2

the seat of secondary hæmorrhage from collateral branches. Some advocate complete division of the vessel before tying, if it be only partially cut across. An artery is tied in the wound, because, in ligating at a distance above, the bleeding might continue through anastomosing branches below the ligature. If the wound is punctured and deep seated, it must be enlarged to enable the surgeon to reach the vessel. The wound at first should only be increased enough to allow the operator's finger to enter and close the bleeding orifice; afterward it can be enlarged at pleasure, without danger. When branches of the external or internal carotids are wounded through the mouth, it is necessary to depart from the rule of ligation in the wound, and tie them in the neck. (*See* article on ligation of arteries.)

*Acupressure*, as a means of suppressing hæmorrhage, has not been before the profession a sufficient time to test its claimed superiority over ligation. The method was first brought into notice by the late Dr. Simpson, of Edinburgh, who, in adopting it, almost abandoned the ligature. The canal of the artery is obliterated by means of a sharp needle introduced at one side of the artery, and passed over the vessel into the tissues on the opposite side. Another method is by passing a common sewing-needle, armed with fine wire, *behind* the artery, allowing the head and point of the needle to be exposed, and then bringing the wire over the tissues covering the artery and fastening it to the point of the needle. In this case the artery is compressed between the wire and needle, and not, as in the former, by the needle and underlying tissues. When acupressure is employed on medium-sized arteries, the needle can be removed in three or four hours with perfect safety, as by that time a firm coagulum will have formed which closes the canal. This method, of course, will give the wound a better chance to heal, as the ordinary ligature is necessarily a source of irritation.

In large vessels, such as the femoral, the needle must remain in for fifty or sixty hours at least, in order to insure success in the operation.

*Cautery.*—Years prior to the introduction of ligation, and before the discovery of chloroform and ether blessed suffering humanity, it was customary, after amputation of a limb, to touch the bleeding stump with a red-hot iron, or to plunge it into a vessel of boiling tar, to stop hæmorrhage. Happily, modern science, in its advances, has driven this barbarous practice almost entirely from the profession. Occasionally, however, even at this day, the actual cautery is made use of. Hæmorrhage from the neck of the uterus, after removal of tumors, and from other organs not accessible to ordinary means, can be thoroughly controlled by the hot iron.

*Position.*—When the bleeding is moderate, simple elevation of the wounded limb will retard the current coming from the heart, and thus assist in stopping the flow. It is always to be employed in conjunction with other measures previously mentioned.

After the cessation of the hæmorrhage, the lost vitality of the patient must be restored. Bottles filled with hot water are to be applied to the extremities, and the body is covered with warm blankets. Tablespoonful-doses of brandy and a few drops of ammonia should be administered every fifteen or twenty minutes, or at such intervals as the case may require, until reaction is thoroughly established. Subsequent irritability of the nervous system is to be treated with opiates.

When the loss of blood is so great that reaction is impossible through the ordinary methods, resort must be had to transfusion. This operation consists in abstracting blood from a robust man or woman, and injecting it into the veins of the exsanguinated patient. If an apparatus for the purpose is not at hand, or its use but little understood, a common hard-rubber syringe, with a capacity of five or six ounces, will answer. An opening is made in one of the veins of the forearm, and into this a canula, adapted to the point of the syringe, is inserted. A bandage tied below the incision prevents further bleeding. The syringe, warmed and charged with the fresh blood, is introduced, and the piston steadily forced down until the instrument is emptied. From ten to twenty ounces may be injected at one sitting, and the operation may be repeated if necessary. Care must be taken to force out all air from the syringe before it is used. The efficacy of this operation has been fully proved. Patients have been restored to life under circumstances which were such as to almost preclude the hope of recovery.

I have lately employed a modification of Dieulafoy's aspirator in transfusion. The arm is bandaged as in the ordinary method for venesection, and a needle of the aspirator inserted into the distended median basilic vein. The stop-cock of the aspirator is then turned, and the blood rushes in and fills up the cylinder. A vein in the patient's arm having been exposed, and an opening made in it for the insertion of a canula, the tube from the opposite side of the aspirator is attached, and the blood forced through it into the vein. See APPENDIX.

# CHAPTER II.

Bleeding from the Nose, Mouth, Lungs, Stomach, Intestines, Kidneys, Ureters, Bladder, Urethra.—Ecchymosis.

EPISTAXIS, or bleeding from the nasal passages, is the most frequent and least dangerous of all internal hæmorrhages. It occurs generally from one nostril. Repeated hæmorrhage from the left nostril is said to be a certain indication of splenic disease.

Some of the capillary vessels of the nasal mucous membrane communicate directly with those of the cranial cavity, and, when epistaxis appears during congestion of the brain, its action is decidedly beneficial in diminishing the quantity of blood in that organ. In inflammations of the mucous membrane, a rupture of the distended and engorged capillaries may be the commencement of a healthy action. All cases of epistaxis, however, are not attended with the same good results: the bleeding may be so persistent as to seriously endanger life.

The ancients considered bleeding from the nose as an indication of fever, and bled and purged the unfortunate patient while any trace of the disorder remained. The blood was supposed to be overheated, and in a state of ebullition, which rendered its removal necessary.

The causes of epistaxis are violent exercise after drinking, laceration of vessels by blows or falls, cardiac disease,
catarrhal inflammations, congestion of the brain, syphilitic
or scrofulous ulceration, the hæmorrhagic diathesis and
disordered conditions of the blood, such as occur in scorbutis, purpura, and continued fevers.

Severe forms of epistaxis are preceded by a feeling of
weight, and fulness about the forehead, with pain and vertigo.

*Treatment.*—First ascertain whether the blood escapes
from both nostrils, or from the right or left; then, on the
affected side, raise the arm above the head, and grasp the
nose with a firm pressure between the thumb and forefinger;
at the same time, a towel saturated with ice-water may be
laid on the forehead. The arm is raised to *distribute* the
force of the heart's action, and to take the pressure off the
carotid vessels, diminishing the strength of the current
through them.

Some advise the application of ice to the mammæ of the
female and testes in the male, or simply placing the hands
in cold water. When pressure, raising the arm, or cold
applications, are unsuccessful, styptics may be resorted to.
Inject with a syringe a quantity of ice-water, or a solution
of common salt, in the proportion of one tablespoonful to
half a tumber of water; or some of the preparations of iron,
such as solutions of the pernitrate or persulphate. The
iron may be thrown up the nostril, either diluted or not, or
a piece of lint, twisted and moistened with the solution, may
be forced up the canal and allowed to remain until the
bleeding ceases. When the blood comes from laceration of
the naso-palatine artery, all these measures are apt to fail.

and the posterior nares must then be plugged. The operation of plugging is simple, and does not require a great amount of skill. It may be performed with Beloc's canula or a gum-elastic catheter (No. 4 or 5 will do). Through the eye of the instrument pass a string, allowing the ends to hang down.

Introduce the catheter through the nostril into the mouth, and draw the string, which is hanging from its end, out beyond the lips. To this attach a piece of sponge sufficiently large to fill up the opening in the posterior nares. Then withdraw the catheter from the nose, and make traction on the string until the sponge is drawn back into the posterior nares, completely filling its cavity. If necessary, the sponge may be dipped in an astringent solution before its introduction. This method scarcely ever fails to control the most obstinate hæmorrhage.

STOMATORRHAGIA.—Hæmorrhage from the mouth. This variety needs scarcely more than a passing notice. It requires special treatment only when occurring in persons with the hæmorrhagic diathesis. Inflammation of the buccal cavity, ulcers, and injuries, are its principal causes. Rinsing the mouth with alum-water, or some other astringent preparation, will check it effectually.

HÆMATEMESIS.—Hæmorrhage from the stomach generally occurs during the progress of some chronic disease of the liver, portal system, or stomach. Any obstruction to the return of blood through the portal vein, such as exists in the dram-drinker's liver (*cirrhosis*), in inflammation or thrombosis of the vein, will occasion it. Chronic ulcer and cancer of the stomach, gastritis, and corrosive poisons, are also prolific causes.

In cirrhosis, the liver is diminished in size by the contraction of new *fibrous* tissue, which is formed throughout
the organ during the inflammatory process.  This new tissue
is either developed from inflammatory lymph (*Rokitansky*),
or by the proliferation of connective-tissue cells (*Virchow*).
It is located principally around the hepatic vessels.  By its
contraction, the ramifications of the portal vein are pressed
upon, and their capacity diminished or destroyed, and the
result is a damming back of the blood in the stomach and
intestines.  In a short time the distention is greater than the
walls of the vessels can resist, and consequently they are
ruptured.  Coagulation of blood in the veins (thrombosis),
with or without inflammation, produces hæmatemesis in a
similar way.

In chronic ulcer and cancer, molecular death of the tissue proceeds gradually, until the capillary walls are reached
and perforated.  If a large vessel have been opened, the
bleeding may cause death in a short period; but such an
event rarely happens.

Instances are recorded of hæmorrhage from the stomach
occurring at the menstrual period.  In this vicarious men-
struation, the usual flow from the uterus is absent.

In profuse hæmorrhage from the stomach, the patient will
have a feeling of fulness and oppression in the epigastrium.
The countenance becomes pallid; there are vertigo and
dimness of vision; and finally a fluid, which imparts a warm
sensation to the œsophagus, is vomited.  If the blood have
been extravasated suddenly and in an empty stomach, there
will be little change in its physical or chemical characteristics.  But if slowly exuded, and allowed to mingle with the
gastric juice, or partially-digested food, it takes on a dark

color resembling "coffee-grounds." The normal alkaline reaction is changed to acid, and the blood will not coagulate. These peculiarities are usually present, and in cirrhosis they are particularly marked. Blood from wounds of the mouth is sometimes swallowed and afterward thrown up, but a careful examination will reveal the source, and prevent an erroneous diagnosis.

The act of vomiting, which forces out the blood in hæmatemesis, is seldom attended with nausea. In passing out some may enter the larynx and induce a fit of coughing, thereby leading to the supposition that the blood is from the lungs, instead of the stomach. On the other hand, a paroxysm of coughing, with hæmorrhage from the lungs, may bring on nausea and vomiting, and cause the physician to locate the disorder in the stomach. It is necessary, therefore, in making a diagnosis, to exercise care and judgment.

It is well to remember that blood from the stomach is generally dark in color, mixed with food, and is acid in reaction. If coagula are present, they will be found black and heavy, from absence of air. There will be a previous history of pain, nausea, vomiting, and a disordered stomach, with the special symptoms of the disease which may have occasioned the hæmatemesis.

In hæmorrhage from the lungs, the blood is generally bright red, frothy, mixed with bubbles of air, and alkaline in reaction. A fit of coughing precedes and accompanies the bleeding. There are pain in the chest, and signs of tuberculosis or other affection of the lungs or cardiac disease, and there is no history of disease of the liver or stomach. Moist râles can be heard on auscultation, near the seat of the pain, and there may also be slight dulness on percussion.

In all doubtful cases, the mouth and fauces should undergo a careful examination. Hæmorrhage from these parts is often mistaken for hæmatemesis. A perfect knowledge of these points of difference, and their careful investigation at the bedside, will make the diagnosis a matter of almost positive certainty.

*Treatment.*—Absolute rest in the recumbent posture must be rigidly enforced in this and every other variety of internal hæmorrhage. The patient's room must be kept free from visitors, and only the nurse and doctor are to be admitted. Every source of excitement must be removed. These stringent preliminaries are, of course, only required when much blood has been lost. There are many mild cases in which they are not called for. Ice stands at the head of all remedial agents for the suppression of hæmatemesis. It can be administered continuously in small pieces, or at different intervals, as the case may demand. Cloths wet with ice-water, or pounded ice in bags, may also be applied over the epigastrium. Ether-spray, directed over the stomach, produces intense cold, and is worthy of trial. Of the various styptics employed, some prefer the following:

R. Liquor. ferri subsulphatis    .    .    .    .    .    3 i.
    Aquæ .    .    .    .    .    .    .    .    .    ℥ ii.   M.

One teaspoonful of this solution is to be given every half hour, or more frequently if required. Other preparations of iron are also used. Some prefer the acetate of lead in one or two grain doses. Alum, creosote, tannic and gallic acids, answer in some cases.

All the solutions employed should be kept on ice, and given in small quantities, as they are apt to be thrown up.

If vomiting is produced by one preparation, let something else be substituted. The contractions of the stomach in the act of vomiting increase hæmorrhage.

The subsequent treatment must depend entirely on the accompanying disease and the amount of blood lost. Nutritious diet and tonics are indicated to restore the lost vitality. When strength is regained, the disease which produced the hæmorrhage should receive special attention. If the bleeding has been so great as to induce collapse, rapid stimulation should be resorted to in the manner described in the preceding chapter.

*Melæna* is a term usually employed to denote hæmorrhage from the bowels, although any dark-colored discharge from the same parts might properly be classed under the same head. Melæna is caused by many of the same disorders which occasion hæmatemesis. The portal venous system, which carries blood from the stomach, also takes it from the intestines. Any abnormal condition, therefore, which obstructs the circulation through the portal vein, such as those previously mentioned, is liable to produce extravasation of blood in any part of the stomach or intestinal canal. Sometimes the blood which is poured out in the stomach passes through the pyloric orifice, and is voided by the bowels instead of being vomited.

Among other causes of bleeding from the intestines may be enumerated ulceration of the mucous coat, from chronic or acute inflammations, and rupture of capillary vessels during inflammatory congestion, as in dysentery and enteritis. *Hæmorrhoids*, or piles, are also classed as common causes. In low forms of fever, such as typhoid or yellow fever, hæmorrhage from the bowels is not of infrequent occurrence. In the first instance, it is due to ulceration; in the second, it arises from rupture of blood-vessels.

When the blood proceeds from the upper part of the intestinal canal, or when it is poured out in small quantities, it appears in dark masses resembling tar. In profuse hæmorrhage it has the same characteristics as when occurring from other organs. When the bleeding is due to ulceration, the blood is generally redder than in rupture of portal capillaries or in piles. Hæmorrhage from intestinal hæmorrhoids (piles) occurs more frequently than any other variety. In cirrhosis of the liver, the gastric vessels are, as a rule, first ruptured, and afterward the vessels farther down the canal. Occasionally, cases of violent hæmorrhage from the bowels, due to cirrhosis, prove fatal in a few moments. Plethoric persons, who feed on the fat of the land, and indulge freely in wine, are at times subject to small hæmorrhages while straining at stool. The portal venous system contains a much larger proportion of fluid during digestion than at any other period, and in plethoric men this distention reaches its maximum, so that, in a violent effort to evacuate the bowels, some of the engorged capillaries rupture and relieve themselves. This variety of melæna occurs independent of any organic disease, not even hæmorrhoids being present to account for it. Hæmorrhage of this character acts as a safety-valve, and should be let alone unless too profuse.

*Treatment.*—The general rules which govern the treatment of other varieties of hæmorrhage must be followed here; perfect rest and quiet secured, and every excitement avoided. Cold water poured slowly from a sprinkler or pitcher is advisable in alarming cases. Cloths wet with ice·

water, or injections of ice-water, or of pounded ice, into the rectum, are beneficial. The vegetable astringents, such as logwood, oak-bark, catechu, tannic and gallic acids, given by the mouth or rectum, act well in mild forms of hæmorrhage. Some prefer the styptic solutions of iron, mentioned in the treatment of hæmatemesis. Small doses of opium, to diminish peristaltic action of the intestines, should always be given. I have found tannic acid and opium, administered by the mouth, and the application of cold water to the abdominal walls, answer admirably in ordinary cases of melæna.

HÆMOPTYSIS.—The occurrence of hæmorrhage from the lungs was at one time considered a certain indication of tubercular deposit. It was a sign of fatal significance in the eyes of physician and patient. A closer investigation of pathological changes in the lung-tissue has demonstrated conclusively the erroneousness of this idea. Hæmoptysis is found, in the majority of cases, to depend on conditions which do not seriously endanger life, and which are amenable to treatment.

The class of persons most subject to this hæmorrhage are those who grow rapidly in height, without a corresponding development in bulk, who are pale and delicate, and subject to common colds and scrofulous inflammations. In these cases there is a general lax condition of the system, a want of tonicity in the capillary vessels, and in other tissues throughout the body, which predispose to hæmorrhage. In inflammation of the larynx, trachea, or bronchial tubes, the vessels of the mucous membrane are distended with blood. A paroxysm of coughing increases the internal pressure on these vessels to such an extent that they rupture, and blood appears in the expectorated fluid. The amount of blood poured out will of course depend on the size and number of the ruptured capillaries. In all cases of catarrhal inflamma tions of the air-passages this rupture and extravasation are li- able to occur, independently of other affections. If the blood were expectorated, the hæmorrhage would be rather a bene- fit than otherwise; but sometimes it remains in the smaller tubes and air-cells, acts as an irritant, sets up inflammation, and finally may go on to consolidation and subsequent soft- ening and degeneration of the lung-tissue (*Niemeyer*).

Organic disease of the heart is accompanied by hæmop- tysis. When insufficiency of the mitral valve exists, the blood regurgitates into the left auricle, which is therefore partially filled with blood that should have remained in the ventricle. This causes a damming back, or obstruction, to the blood coming from the four pulmonary veins to the auricle, and consequent congestion of the lungs. The capil- lary vessels in the bronchial tubes, and in other parts, are distended, and relieve themselves by rupture.

Sometimes, in these cases, large extravasations of blood occur in the parenchyma of the lung (*pulmonary apoplexy*), lacerating and destroying its substance, and hastening a fatal termination. Extravasations of blood in cardiac dis- ease are also due to another cause, viz., the plugging of small arterial capillaries by clots of fibrine detached from the right side of the heart. These clots are carried into the pulmonary artery, blocking up some of its terminal branches. This obstruction necessarily diminishes the current in the capillaries supplied by the plugged vessel; they become crowded, choked up with blood, the internal pressure soon forces their thin walls to give way, and the blood is extravasated into the air-cells, terminal bronchi, and between the elastic fibres of the cells. These clots, after coagulation, are circumscribed, sharply defined, and dark in color. To this old condition a new name has been given, viz., *hæmorrhagic infarction*, to distinguish it from another variety of pulmonary apoplexy in which the clot is diffused, and lungtissue destroyed.

Tubercular deposit induces hæmoptysis in one of three ways: 1. By mechanical pressure it may obstruct the small attenuated vessels so as to cause rupture; 2. It may create inflammatory congestion, which is relieved by the walls giving way; or, 3. The softening and degeneration of tissue which accompany the second and third stages of tuberculosis, involve the capillaries, destroy them, and hæmorrhage is the result.

Gangrene of the lung is seldom accompanied by hæmoptysis. When present, it is due to the morbid process including the vessels in the general destruction.

The hæmoptysis which occasions the characteristic rustcolored sputa of pneumonia either arises from laceration of the minute capillaries, or by the passage of the red globules through the wall of the vessel without rupture. The latter process is doubtful, to say the least of it.

The inhalation of chlorine gas, sulphuretted hydrogen, and other irritating substances, likewise occasions hæmoptysis. Wounds of the lung are always attended by more or less expectoration of blood.

One curious and rare variety of hæmoptysis is that which occurs at the menstrual period, when the discharge of blood from the uterus is absent. There are but few cases on record. Dr. Watson relates one of a young girl who menstruated once naturally at sixteen years of age, and, from that time until the age of fifty, she suffered from hæmoptysis regularly once each month. Accompanying the loss of blood were the usual uneasy sensations of pain in the pelvis and general malaise.

In slight cases of hæmoptysis the patient has first a tickling sensation, beneath the sternum, which compels him to cough. The effort brings up a warm fluid having a peculiar sweetish taste, which when expectorated is found to be blood. It is generally bright red, and filled with bubbles of air. At other times the sputa for some days are simply tinged or streaked with red. In more serious cases, and especially in heart-disease, there is a sharp, intense pain in some part of the chest, followed immediately by excessive dyspnœa, and the expectoration of large quantities of blood. This blood is not so bright as in the former instance, but it still contains air. On auscultation near the seat of extravasation, moist râles, and occasionally bronchial breathing, can be heard. The râles are more liquid in character than those produced by mucus. There is more or less dulness on percussion, in the majority of cases. These large extravasations are usually followed by pneumonia. Its advent is easily recognized by the characteristic physical signs, and by the increased temperature, rapid pulse, and other evidences of febrile excitement.

In examining a case of supposed hæmoptysis, it is well always to take into consideration the fears of the patient, when determining the quantity of blood lost. The fright causes the amount to be greatly exaggerated. Investigate carefully the condition of the nose, mouth, and fauces. Blood from these parts may get into the larynx, excite coughing, and be expectorated, thus leading to an erroneous diagnosis. The differentiation between hæmoptysis and hæmatemesis is readily made. In the latter the blood is dark-colored, acid in reaction, uncoagulable, does not contain air, and is expelled by the act of vomiting. With it there is a history of some disorder of the stomach or liver. In the former the blood as a rule is red—it is alkaline in reaction, coagulable, filled with bubbles of air, is brought up by coughing, and there is a previous history of some variety of lung-disease (*see* Hæmatemesis).

*Treatment.*—The patient should be placed in a sitting posture in bed, propped up with pillows. A cool room is desirable. Every cause of excitement must be removed. The variety of medication demanded depends to a certain extent on the cause of the hæmorrhage. If it be due to cardiac disease, and if the heart's movements be accelerated, it will, of course, be expedient to administer an arterial sedative in conjunction with the astringent. For this purpose the following prescription will be found of service :

℞. Ext. verat. viridis . . . . . . fl. ℥ ss.
     Ext. ergotæ . . . . . . . fl. ℥ ij.
     Acidi sulph. aromat. . . . . . ℥ ij.
     Aquæ . . . . . . . . fl. ℥ ij. M.

Administered in 30-drop doses, largely diluted, every half-hour, until the desired effect is produced. Digitalis may be substituted for veratrum, or given separately. Great care must be exercised in its administration. For the urgent dyspnœa, which also accompanies this hæmorrhage in heart-disease, the application of half a dozen dry cups to the thorax will be found an admirable remedy. They relieve the troublesome shortness of breath, and, by drawing blood to the surface, diminish the congestion of the lungs.

If there be no special contraindication, the following preparation of sugar of lead and opium, although incompatible, will often answer the purpose :

℞. Plumbi acetatis . . . . . . . ℨ s.
Pulv. opii . . . . . . . . gr. ij. M.

Make ten pills. One to be given every half-hour. In simple cases, one of the oldest, and, at the same time, one of the best, remedies is common salt, alone or with vinegar. Half a teaspoonful can be given at intervals of fifteen minutes until the hæmorrhage is controlled.

℞. Acidi sulph. dil. . . . . . . . fl. ℨ ij.
Aluminis . . . . . . . . . ℨ j.
Aquæ . . . . . . . . . fl. ℥ ij. M.

Can be taken in teaspoonful doses every half-hour. Some prefer the preparations of iron. Inhalation of the vapor of tr. ferri chloridi has been recommended, but its irritating properties would tend to excite coughing, and therefore should not be employed. Gallic acid in three-grain doses, and other vegetable astringents, are found efficacious. In connection with the internal remedies mentioned, hot applications to the dorsal region of the spinal column, and cold ones in front, will be of service. When all danger from loss of blood has passed away, the disease which produced it, and the inflammation (if any) which follows, should receive careful attention.

HÆMATURIA.—Blood in the urine is a symptom of many varied pathological conditions distinct in character and in location. Having its origin in different organs some considerable distance apart, a correct appreciation of its source is attended with greater difficulty than are hæmorrhages from the viscera. Lesions in any part of the genito-urinary tract from the kidneys, ureter, bladder, prostate gland, or urethra, may bring on hæmaturia.

Constitutional blood-diseases, as purpura, scurvy, typhus or yellow fever, are classed as causes independent of special disorders in the organs mentioned.

Hæmorrhage from the kidneys arises from external violence, inflammation of the tubes or parenchyma of the organ ; the passage of renal calculi, or ulceration resulting from the infarction of these bodies, in or near the pelvis. The passage of large calculi through the ureter tears the mucous membrane, and bleeding results.

Blood is found in the urine in injuries of the bladder from introduction of instruments or blows on the hypogastrium, acute cystitis, fungous degeneration of the mucous membrane, and cancerous disease of the organ. Urethritis, chordæ, and injuries of various kinds, are prolific causes of hæmorrhage from the urethra. Various medicinal agents, such as cantharides, turpentine, etc., etc., given in overdoses, produce excessive congestion in the genito-urinary tract which is often accompanied by hæmaturia.

When called to a case of supposed hæmaturia, it will be well first to determine whether blood is present in the urine or not, and then endeavor to discover its source. Healthy urine is a clear "amber-colored fluid," acid in reaction, and having a specific gravity ranging from 1.118 to 1.125. Urine which contains blood has a smoky tint, if the quantity be small; dark red or chocolate-brown, when the quan-

tity is large.   The reaction in most cases is alkaline, and the
specific gravity is increased.   On being allowed to stand, a
dark-reddish mass sinks to the bottom, while the superna-
tant fluid still maintains, to a certain extent, its smoky
hue.   Heating the liquid will give a cloudy precipitate of
albumen, tinged with the coloring matters of the blood,
while the rest of the urine remains clear.   The surest
method of diagnosis is by microscopical examination.
Blood-corpuscles are recognized by their "yellow color,
uniform size and non-granular surface" (*Bird*).

There are many substances besides blood which give a
reddish color to the urine.   An excess of urates in otherwise normal urine will induce a red or brown deposit when
the liquid cools.   To determine their presence apply heat,
and the urine will resume its natural transparency.

The use of beet-root, madder, logwood, etc., also occasions a red color.   The applications of heat in these cases
will not produce a precipitate, showing that the tinge is not
due to blood.

When the blood proceeds from the kidneys, it will be,
generally, diffused throughout the urine.   It will be attended with a history of injury, the passage of a calculus, or
signs of nephritic inflammation.   A microscopical investigation will show small blood-casts of the uriniferous tubules,
red globules, and epithelium from the pelvis of the kidney.
If the blood come from the commencement of the ureter,
small plugs of fibrine, resembling maggots, may sometimes
be seen in the bottom of the glass.

In hæmorrhage from the bladder, more blood comes
away at the end of micturition than during the act; it is
clotted, and not diffused through the liquid, as in the former

instance. There is a history of injury, signs of cystitis, such as frequent desire to micturate, pain during the act, and pain on pressure over the pubes, or signs of stone.

When the bleeding takes place from the urethra, the blood precedes the stream of urine. There is one exception to this rule, namely, where partially-healed ulcers exist in the canal. The contraction of the urethral walls, as the last drops of urine pass out, lacerates some of the delicate vessels in the ulcer. I have known this to occur in several instances.

A careful consideration of the foregoing points of difference will, in most cases, enable the practitioner to make a correct diagnosis.

*Treatment.*—When injury or disease of the kidney causes hæmorrhage, little treatment is necessary, except that which is calculated to remove the existing morbid condition of the organ. In hæmorrhage from the bladder the cause is different. Profuse bleeding from this organ is not infrequent in malignant disease, or fungous degeneration of the mucous membrane. The patient should be placed on his back, and cold wet cloths applied over the hypogastric region and perinæum. Ice-water, or pounded ice, can be thrown into the rectum at the same time. Should the bladder be distended with clots, a large-sized catheter must be introduced, the clots broken up and removed; warm water injected through it will soften the clots and assist in their discharge. If further measures be necessary to suppress the bleeding, the following solutions may be injected into the bladder, by means of the catheter:

℞. Acidi gallici . . . . . . . . ℨ iij.
Aquæ . . . . . . . . . fl. ℥ iv. M.

Or,

℞. Aluminis . . . . . . . . . . ℥j.
Aquæ . . . . . . . . . . . fl. ℥ iv. M.

Many of the vegetable astringents, as uva ursi, hydras-
tis, krameria, may be used in a like manner.

In urethral bleeding, cold cloths and pressure generally
answer all requirements. If there be laceration of the erectile tissue surrounding the urethra, accompanied by dangerous hæmorrhage, a steel sound, or catheter, must be introduced in the canal, and the penis bandaged over it firmly.
This procedure is allowable in every case which cannot be
controlled by other means. In case injections into the urethra are considered advisable, solutions of iron may be employed diluted, such as—

℞. Liquoris ferri subsulphatis . . . . . . fl ℥j.
Aquæ . . . . . . . . . fl ℥ iv. M.

Any thing stronger than this creates much irritation and
pain.

After amputation of the penis, or the removal of tumors,
the subsequent hæmorrhage from the erectile tissue is sometimes so profuse and uncontrollable by ordinary means as
to compel the surgeon to apply the actual cautery. See Ap-
PENDIX.

ECCHYMOSIS is an extravasation of blood in the meshes
of the cellular tissue, generally occurring underneath the
integument. It is especially apt to take place in those parts
which are loosely attached to the underlying tissues, and
where there is little subcutaneous fat. A characteristic example of this lesion is found in the ordinary "black eye."

Ecchymosis follows blows and contusions of all kinds. Its extent depends on the tissue bruised, and the amount and kind of violence which produced it. Very slight injury will occasion large ecchymosis in old persons, and in those who suffer from anæmia or other debilitating affections. In purpura and scorbutis, blood is effused in small, irregular patches. This is due to deterioration of the circulating fluid, and not to injury. The ecchymosed spot may be black, green, yellow, or crimson. Sometimes there is a mixture, the central part being dark blue, while the rest varies in color from a crimson to light green and yellow. The coloration is due to the red globules which have escaped from the ruptured capillaries, and to the hematine of the blood staining the parts. Where the staining is caused by hematine alone, the colors are light, and microscopical examination of the extravasated material shows that no corpuscles are present.

All bruises which are not attended with grave destruction of tissue may be treated with water-dressings. The injured part is to be kept at rest and covered with cold, wet cloths. If preferred, the bruised tissue may be bathed or kept moist with the following preparation :

℞. Ammoniæ muriat. . . . . . . ℨj.
　Tinct. arnicæ . . . . . . . . fl. ℥j.
　Spts. vin. rect. . . . . . . fl. ℥ij.
　Aquæ . . . . . . . . fl. ℥ iij. M.

For children, a further dilution is necessary, as their integumental covering is much more delicate than that of adults. One or two ounces of water added will weaken it sufficiently. This solution has an admirable effect in producing rapid absorption of the effused material, preventing inflammation and excessive discoloration. If there be much pain, the officinal lead and opium wash will give relief. A large extravasation of blood should be removed by incising the integument.

# CHAPTER III.

Metrorrhagia. — Accidental Hæmorrhage. — Placentia Prævia. — Post-partum
Hæmorrhage.

THE periodical discharge of blood from the uterus, which takes place every twenty-eight days, is a physiological occurrence, and does not require attention here. It rarely calls for active treatment, even when in excess (*menorrhagia*).

METRORRHAGIA, or bleeding between the monthly periods, may keep up so constant a drain on the system as to destroy by exhaustion, or predispose to fatal diseases. Congestion of the uterus from chronic inflammation, tumors, ulcers, and abrasions of the cervix, are its principal causes.

The treatment of metrorrhagia consists principally in the application of cold to the hypogastrium, vulva, and neck of the uterus, and the internal administration of ergot, gallic acid, acetate of lead, etc. India-rubber bags, filled with ice-water, introduced into the vagina and pressed against the cervix uteri, may be used with good effect. The diseases causing the hæmorrhage should subsequently be removed, and the patient's strength increased by fresh air, exercise, good diet, and tonics.

ANTE-PARTUM HÆMORRHAGE is that variety which occurs

in the pregnant female before delivery. It is due either to partial separation of the after-birth from blows or falls (*accidental hæmorrhage*), or to placenta prævia. In the latter case, the after-birth is attached around the os internum. The natural dilatation of the cervix and contraction of the uterine fibres at "full term" cause its detachment, and bleeding follows (*unavoidable hæmorrhage*). Placenta prævia is attended with great danger, both to mother and child. It requires to be diagnosed from accidental hæmorrhage. In accidental hæmorrhage, the patient has received a blow or fall on the abdomen, the cervix is not relaxed, and the flow of blood occurs between the uterine contractions. In unavoidable hæmorrhage, the bleeding appears near the time of labor, and is *not* accompanied by a history of injury. The cervix is soft and patulous, the placenta can be felt over the internal os, and the hæmorrhage occurs *with*, and not between, the uterine contractions, as in the former variety.

A patient suffering from accidental hæmorrhage should be kept at rest in the recumbent posture, with the hips elevated. Cold may be applied to the vulva, and astringent medicines given. Some advise small doses of ergot. If these measures do not succeed, premature labor must be induced and the uterus emptied (*see* Puerperal Convulsions).

PLACENTA PRÆVIA is treated in one of four ways: 1. The vagina can be *tamponed*, and the patient kept quiet until labor sets in. The placenta is then removed, totally, and the child's head, pressing against the open vessels, prevents further loss of blood. 2. If the hæmorrhage is profuse, the cervix may be dilated rapidly, the placenta detached as in the first instance, and the child extracted by means of forceps or version. 3. The after-birth may be partially detached at one side when the os is dilated, and the child delivered by version. 4. An opening may be made in the centre of the placenta, the hand introduced through it, and version performed.

Ergot should be freely administered while the uterus is being emptied. This drug is likewise useful after completion of delivery, in producing perfect tonic contractions of the uterine muscular fibres, and preventing further bleeding.

Post-partum Hæmorrhage is one of the most dangerous sequelæ of labor. Perhaps in no other hæmorrhage is there such urgent necessity for presence of mind, or active interference. There are few varieties which so readily yield to proper treatment; yet inferior remedial agents, or a few moments of indecision, may place the patient beyond hope. The stream of blood poured out in the space of half a minute has in some instances been sufficient to destroy life.

Protracted labors which fatigue and lessen the vital forces of the parturient woman, or labors which have been attended by operative procedures, are apt to be followed by profuse bleeding. Neglect on the part of the physician or of his assistant to follow the uterus with the hand down into the pelvis during delivery, and to keep it contracted when there, is one of the most common causes. It is not too much to say that, if this precaution were observed with all patients, a case of immediate post-partum hæmorrhage would be exceedingly rare.

Women habitually subject to inertia uteri are especially liable, even in ordinary labors, to lose large quantities of blood. These cases require extra attention. Injuries to any part of the internal genitals, with laceration, and the

hæmorrhagic diathesis, are also causes of immediate hæmorrhage.

When portions of the after-birth remain behind after
delivery of the child, hæmorrhage usually occurs. It does
not, however, show itself to any great extent for some days
subsequent to the labor. Retained placenta may be suspected in all cases where a few days elapse after delivery before the bleeding manifests itself.

In post-partum hæmorrhage the blood may be effused
into the cavity of the uterus, or, as is generally the case, it
may be poured out through the vagina.

The first indication of hæmorrhage which may attract
the attention of the attendant, especially if the woman be
covered or the bleeding internal, will be a sudden blanching
or pallor of the patient's countenance, and sighing respiration. The pulse becomes rapid and weak, or may be completely absent. In short, all the constitutional symptoms of
profuse hæmorrhage are present (*see* page 12). In another
class of cases the bleeding is slower, the constitutional
effects less suddenly manifested; but in all they appear to a
greater or less degree.

*Treatment.*—The preventive treatment consists in pressing the uterus firmly down into the pelvic cavity as it is
being emptied of its contents, and to keep the hand over it
until it is felt to be contracted like a hard ball in the pelvic
cavity. Some recommend the administration of ergot before
and after the delivery of the placenta, as a preventive measure. I administered it quite frequently for that purpose
in the Lying-in Department of Bellevue Hospital, and with
good results.

For suppressing the hæmorrhage, several methods are

advised. When the bleeding is very profuse, the surest method is to introduce one hand into the uterus, turning out all the clots, while at the same time the other hand grasps the organ on the outside, and firm pressure is made until the hand is forced out by the uterine contractions. A piece of ice may be carried into the cavity, and applied to the internal surface of the uterus, if necessary. The physician must be governed by circumstances in its use. There are cases which cannot be controlled without it. Some object to the introduction of the hand into the uterus, because they think it apt to injure the walls, produce endo-metritis and other disorders. This danger is probably somewhat exaggerated. The pressure of the closed hand for a few moments on the inner surface of the contracting uterus will certainly not produce greater harm than the pressure on the irregular prominences of the child's body during a labor of several hours' duration. The only danger there can be is from septic material finding its way inside on the hands of the physician, and this, to say the least, is very improbable.

Another method is to grasp the uterus firmly and knead it with the fingers until contractions ensue. Lumps of ice may be rubbed over the abdomen at the same time, or ice-water poured on the abdominal walls.

Prof. Thompson, of this city, claims to have obtained good results from the application of ether-spray over the hypogastrium. Injections of astringent medicines into the cavity of the uterus have been employed, but are considered extremely dangerous by most obstetricians. In conjunction with all the varieties of local treatment mentioned, ergot should be administered in large doses at repeated intervals. Its use is always indicated. The subsequent treatment

depends on the amount of blood lost. If there be much
exhaustion, the usual stimulants, together with small doses
of opium, may be given; and, as a last resort to save from im-
pending death, the operation of transfusion, referred to in a
former chapter, may be employed. Injections of hot water
have also been employed with great advantage.

# CHAPTER IV.

Wounds of the Throat, Lungs, Pericardium, Heart, Abdomen, Intestines, Bladder, Perinæum, Joints, Arteries, Veins.—Perineal Section.—Paracentesis, Thoracis.—Gunshot Wounds, etc.

WOUNDS of the throat vary in extent, from simple incision of the integument to complete severance of the larynx, trachea, and œsophagus. They are inflicted with razors or other sharp cutting instruments, and are usually the result of attempted self-murder. The upper part of the throat seems to be the point of selection in these cases: rarely is the cut made at the lower portion. The carotid artery and jugular vein are thus saved, and a better chance of recovery given to the patient.

In the majority of wounds of the throat an opening is made into the air-passages. The most common seat of these wounds is between the thyroid cartilage and hyoid bone, and over the larynx. In the former the thyro-hyoid membrane is cut through; the epiglottis may be cut off, or injured so as to seriously affect the power of swallowing. The food may pass without hinderance into the larynx and out of the external opening, as the epiglottis is not in place to prevent it, or is in a semi-paralytic condition from the injury, and fails to appreciate, or prevent the passage of the food down the wrong canal. The appearance of food in the wound is therefore not a positive indication of injury to the œsophagus.

Wounds inflicted on the side of the neck may cut the pneumogastric or phrenic nerves. In such cases there is interference with the respiratory movements, and subsequent congestion of the lungs, which may ultimately destroy life, independent of any other complications. Wounds of the back of the neck, unless implicating the spinal cord, are not fatal. Some authorities say that they are followed by paralysis of the lower limbs and loss of sexual power; this is doubtful.

Wounds inflicted between the lower jaw and hyoid bone are the least dangerous of anterior wounds, although they are sometimes attended with great hæmorrhage and with difficulty in swallowing (dysphagia).

The danger and causes of death in wounds of the throat are: 1. Hæmorrhage; 2. Asphyxia. 3. Inflammation of the air-passages and lungs, as laryngitis, bronchitis, and pneumonia. 4. Nervous depression and starvation.

The principal danger is from excessive bleeding. Bleeding may be profuse even in superficial wounds. The blood from the numerous plexuses of veins in front of the neck and around the thyroid gland may flow in sufficient quantity to destroy life. When the large vessels, such as the carotid arteries or jugular veins, are cut, death occurs in a few moments.

Secondary hæmorrhage not unfrequently takes place from sloughing of the walls of the vessels, between the tenth and the twentieth day.

Asphyxia may arise from infiltration of serum into the mucous membrane of the larynx at its upper part (*œdema glottidis*), or from blood flowing down into the air-passages. Internal hæmorrhage may go on slowly for some time without attracting special attention, the shock of the injury and deficient aëration of the blood benumbing the sensibility of the mucous membrane.

Laryngitis may occur from extension of inflammation from surrounding parts, or directly from a wound of the larynx. The most dangerous inflammations are bronchitis and pneumonia. These complications arise principally from the inhalation of cold air through the opening in the throat. In ordinary breathing, the air is heated by passing through the nose, and thus loses its irritating qualities.

In all suicidal attempts upon life, there is extreme mental depression, which tends to prevent recovery.

*Treatment.*—As the great danger arises from loss of blood, the first efforts are directed to suppress the flow. This is accomplished either by means of *pressure*, or with the *ligature*. If the bleeding vessel cannot be reached in the wound, sufficient pressure may be made to stop the hæmorrhage, while the upper or lower portions of the wound are enlarged and the vessel searched for. Should it not be found, and the hæmorrhage be still threatening, the carotid arteries must be tied. If the wound does not implicate the air-passages, the edges may be drawn together with strips of adhesive plaster. In doing this, care should be taken to leave an opening for the discharges from the wound. The cellular tissue of the neck is very loose, and, unless this be done, pus and other inflammatory products will burrow at the base of the neck, between the muscles and vessels, and produce serious trouble. The same rule holds good when the wound extends into the air-passages. No attempt

4

should be made to close the aperture for several hours, or until all danger from hæmorrhage has passed away. Even then the central portion of the wound should remain unclosed for the exit of the subsequent discharges. In closing the wound and preventing gaping, the head should be flexed on the neck, and retained there by means of bandages passed over the head and under the arms. Cloths wet with cold water may then be applied to lessen inflammation. If there is venous oozing in the canal, a large tube may be introduced, and pressure made by plugging around it (*Ericcson*).

When the œsophagus is wounded, the patient can be fed through the opening by means of a flexible catheter, or the tube of an ordinary stomach-pump. I have found the latter to be much better for the purpose than the catheter, as a larger quantity of food can be introduced in a given space of time, and the wound therefore sooner relieved from the presence of an irritating substance.

Patients should always be removed to a very warm room, with a temperature of from 80 to 85° Fahr. Stimulants, and nourishing diet, in the shape of beef-tea or chicken-broth, should be freely administered.

WOUNDS OF THE THORAX, LUNGS, ETC.—Non-penetrating wounds of the thorax are treated like simple wounds in other parts of the body. They do not require consideration here.

Penetrating wounds may involve the internal mammary and intercostal arteries, the pleura, lungs, heart, and great vessels, either alone or collectively. When the internal mammary artery is cut, the blood flows slowly into the anterior mediastinum, or into one or the other pleural cavities. It is diagnosed by the location of the wound and the gradual development of syncope consequent upon the loss of blood.

The protection afforded to the intercostal vessels, by tho long groove in which they run, happily prevents them from being wounded, except in very rare instances. In wounds of these vessels, the hæmorrhage may take place in the cavities of the pleura, underneath the muscles and fascia of the chest, or escape internally. The immediate danger to life is not very great, but the utmost difficulty in suppressing the hæmorrhage is commonly experienced.

Penetrating wounds of the chest, without injury to the lungs, are exceptional. Injury to the lungs may be excluded, if there is no expectoration of blood, or hæmorrhage from the wound. If the hole is large, sufficient air may pass into the cavity of the pleura to compress the lung and completely destroy its action. In such a case, death may ensue.

The most dangerous wounds of the lung are produced by bullets. Foreign bodies in the delicate structures of the lung cause great irritation, and more inflammation than simple laceration would. They are not, however, necessarily fatal. Many instances are on record of foreign bodies remaining embedded in the lung-substance for years, without interfering specially with respiration. In the summer of 1868, I made a *post-mortem* examination on the body of Major D——, an old Mexican veteran who had received a gunshot-wound twenty years before. In the upper portion of the left lung was embedded a large, old-fashioned musket-bullet, completely encysted. The lung was about one-quarter its original size, and was carnified around the projectile. The major had enjoyed comparatively good health, notwithstand·

ing its presence. He, strangely enough, supposed that the bullet was in the lung of the opposite side, and his friends were of the same opinion.

The signs of a wound of the lung are plain and well marked. There is great difficulty in breathing (*dyspnœa*), expectoration of blood (*hæmoptysis*), and of red, frothy mucus from the air-passages, and emphysema. There may or may not be hæmorrhage from the external opening. On auscultation, small moist râles may be heard near the seat of injury. The patient's face is pallid and anxious, and the pulse small and rapid. In some cases the bleeding goes on inside the chest, until the lung is compressed by it, and signs of syncope show themselves. Internal hæmorrhage may be diagnosed by the increased paleness of the countenance, flickering pulse, vertigo, and dimness of vision, increased dulness over the affected side, absence of the respiratory murmur. If the blood be poured out to any extent in the parenchyma of the lung, there will be dulness on percussion near the wound, and bronchial breathing.

The passage of air into the cellular tissue (*emphysema*) is a common accompaniment of wounds of the lung. It may occur when a part of the lung-tissue is ruptured by pressure on the chest-walls, or penetrated by the broken end of a rib, independent of any external wound. When it proceeds from rupture of the vesicles alone, and extends to the surface, its usual course is through the cellular tissue of the posterior mediastinum up to the neck, whence it travels to other parts of the body. A case of this kind came under my care in Bellevue Hospital, in a patient whose chest had been severely injured by a derrick. The ribs were not, however, broken. In a few hours after admission to the ward, emphysema manifested itself, and spread slowly over the neck and face, and finally involved the thorax and abdomen. The face, arms, and trunk, became distended to an extreme degree. He suffered greatly from pain and difficult respiration. There was some expectoration of a reddish-colored, tenacious mucus, circumscribed bronchial breathing over the left lung, near the apex, a hot skin and rapid pulse, with other indications of pneumonic inflammation. It was regarded as a hopeless case. In ten days from the time of admission, the emphysema diminished rapidly, and, at the end of three weeks, no trace of it was present. The patient was discharged cured.

In wounds which open externally, the air is drawn in with each inspiration, and forced out during expiration, some of it passing into the cellular tissue. It may remain localized near the wound, or it may extend gradually to other parts. Emphysema is always recognized by the elasticity of the swelling, and by the peculiar crackling, crepitant sensation, communicated to the fingers on pressure.

The air, instead of passing out into the cellular tissue, may accumulate in the pleural cavity, giving rise to *pneumothorax*. In certain cases of hæmorrhage, this has a salutary rather than an injurious effect, as the compression of the lungs will stop the flow of blood.

PNEUMOCELE, or hernia of the lung, may take place before the external wound heals, or after it is entirely closed. When protruded through the wound, it may be pushed partly back, and the aperture closed by a compress. Some cases of pneumocele have been treated by cutting, and by strangulating the extruded portion. If the hernia be a remote result of the wound, and covered by the integument, all that is necessary is to protect it by a hollow pad.

*Treatment.*—When the intercostal arteries are wounded, they may be either compressed or ligated. Ligation is almost impossible. The best method is to fasten a piece of sponge to a ligature and force it through the wound into the cavity of the chest, and then draw it partially outward so as to make it press directly upon the arteries (*Poland*). Digital compression, kept up by relays of assistants, has in some cases been effectual. Some recommend passing a silk or wire ligature around the rib, drawing tightly, and thus closing the wounded vessel. See APPENDIX.

Others close the external wound, and allow the blood to escape into the cavity of the chest. A large quantity of blood may be lost in this way, but not enough to destroy life.

Wounds of the internal mammary arteries are more difficult to reach than the preceding. Pressure may be tried, in the manner described above. If it do not succeed, ligation may be resorted to. This operation is usually performed at some point above the fourth interspace; below this point, the operation cannot succeed.

The method of ligating the artery is described by Dr. Poland * as follows: "An incision is made two inches in length along the side of the sternum, and in an oblique direction, from above downward, and from without inward, forming with the axis of the body an angle of forty-five degrees: the centre of the incision to be three or four lines from the border of the sternum.

"Having divided the skin, cellular tissue, and origin of the pectoralis major muscle, the intercostal space is brought

* Holmes's Surgery, article Wounds.

into view; the intercostal muscle is now carefully divided upon a director, and the edge drawn apart by retractors, and the arteries exposed."

In WOUNDS OF THE LUNG an attempt must be made to control the hæmorrhage by internal medication. Small doses of acetate of lead, sulphuric acid, alum, or other astringents, may be given. Ice applied externally is always of service. Should the blood accumulate in the interior, it must be removed. If it does not flow out by changing the position of the patient, a cupping-glass may be placed over the aperture, and the fluid started in this way. Of course, this procedure should not be instituted while any danger of further hæmorrhage remains. Some prefer enlarging the external wound, while others allow it to heal, and afterward perform *paracentesis thoracis.*

This operation is usually made posteriorly near the angle of the scapula, between the seventh and eighth ribs. The best instrument to employ is a small trochar and canula. When the point of opening is selected, the integument is incised with a scalpel, and the trochar introduced. As the stylet is withdrawn, the patient should be turned over on the affected side, and firm pressure made on the thoracic walls. In this way there is little danger of air entering the cavity. Dr. Bowditch, of Boston, uses a suction apparatus to prevent air from passing in, and to assist in evacuating the liquid. Dieulafoy's aspirator answers all purposes.

When the hæmorrhage has ceased, the external wound is thoroughly closed, and the lips held together by adhesive plaster. Simple water-dressings, dipped in a solution of carbolic acid, are then applied over the part until it is healed.

If pneumo-thorax exist of sufficient extent to compress

the lung, the enclosed air may be extracted by suction, through the external wound, or by making a new puncture in the chest-walls.

The subsequent inflammation of the lung-tissue is treated by counter-irritation over the chest, diaphoretics, anodynes, etc.

WOUNDS OF THE PERICARDIUM.—A punctured wound in the præcordial region, which does not implicate the heart or great vessels, is of rare occurrence. Such a wound may prove fatal from the entrance of blood or air into the peri cardial sac, pressing upon the heart so as to paralyze its movements. The inflammation of the pericardium which follows a wound of this kind may also destroy life.

This wound is recognized by the ordinary signs of peri- carditis. Upon auscultation there is heard a dry, rubbing friction-sound accompanying the cardiac impulses. This is succeeded by an augmentation of the area of præcordial dulness from effusion, and by diminished intensity of the heart-sounds, and feeble pulsations. The constitutional effects are shown by a rapid, irritable pulse, hot skin, and anxious face.

When the hæmorrhage has been controlled, the wound is closed in the ordinary way, and opium is administered in full doses. Blisters, dry or wet cups over the præcordia, are effective agents in subduing the inflammation.

WOUNDS OF THE HEART may be instantaneously fatal, or life may be prolonged for several days. The case of a noted pugilist of this city, named Poole, will be remembered. He received a bullet-wound in the heart, and walked home afterward. Death did not occur for hours after the injury was inflicted.

Small punctured wounds of the heart have been known to terminate in recovery.

A wound of the auricles is more rapidly fatal than a wound of the ventricles. The walls of the former are thinner, and the fibres more uniformly arranged, and their contractions less likely to prevent hæmorrhage. The muscular walls of the ventricles are thick, and the fibres interlaced, and, if the wound be small, profuse bleeding cannot occur.

The signs of wounds of the heart are those of shock and loss of blood. The patient becomes rapidly insensible, and the pulse ceases. There is extreme pallor. The extremities are cold and sometimes clammy. When the immediate danger has passed, signs of pericarditis appear. If life be prolonged sufficiently to give chance for treatment, the patient is to be kept perfectly quiet, the wound closed, and covered with cold-water dressings. Opium is given internally, and, when inflammation appears, remedies are employed as in the preceding case.

WOUNDS OF THE ABDOMEN OR ITS CONTENTS.—Penetrating wounds of the abdomen are generally either punctured or incised. Lacerated wounds are not frequent. If the bleeding is in any way profuse, the vessels should be tied. If the wound is small, so as to make it impossible to reach the vessel, the opening must be enlarged to make it accessible. When there is simply an oozing from the wound, interference is not necessary. It is better for the blood to escape outside than into the peritoneal cavity. The great danger in these cases, as in all wounds of the abdomen or its contents, is peritonitis. This dreaded complication is made known by the occurrence of a sharp pain near the wound, which soon extends over the whole abdomen. There are also tympanitis, constipation, and vomiting. The pulse is hard, tense, and wiry. The skin is dry and the temperature increased. When the intestines are wounded, there is still greater liability to peritonitis. If the opening is large, there is always an escape of fecal matter into the peritoneal cavity. This irritating material is certain to excite peritonitis, even when in minute quantities. A small wound of the intestines may be closed by eversion of the mucous membrane.

*Treatment.*—If the intestines protrude externally, and cannot easily be returned through the wound, the opening should be enlarged. The intestine should be cleansed thoroughly in tepid water before it is returned. If the intestinal wound is more than three or four lines in length, its edges should be drawn together by means of sutures. An opening, of such a size as to be completely closed by the everted lining membrane, may be let alone. Ericceon recommends passing a ligature around this variety, in order to make the escape of fecal matter an impossibility.

In dealing with wounds of the abdominal wall, there is some discrepancy of opinion. Some believe that the sutures should merely include the skin, and not the deeper structure below. It is reasonable to suppose that, in closing the wound in this way, a separation to a greater or less extent would take place in that portion below the integument. Inflammatory products must fill up the gap, and there is nothing to prevent their getting into the peritoneal cavity and giving rise to peritonitis. Unless there are special indications to prevent it, it is better to pass the needle down to the peritonæum, and bring all parts of the wound in complete apposition. If there is much suppuration following the wound, it should be opened, kept clean with carbolic-acid wash, and free escape of pus allowed.

Opium is given internally to control the inflammation and allay pain. The patient should be brought under its influence until his respirations are down to 14, and his skin perspiring. Light poppy fomentations are also of much benefit.

Contusion of the abdominal walls may lacerate the integument or muscles, and the viscera within. The internal organs alone may be injured, without any perceptible lesion of the walls. Severe contusions are scarcely ever recovered from. As a good example of the manner in which these wounds are received, and their mode of termination, the following case may be of interest:

James D., aged twenty-seven; occupation, laborer; was admitted to Ward 11, Bellevue Hospital, suffering from a severe contusion of the abdomen. He had been riding on the rear platform of a Third-Avenue car, which was driven at considerable speed. The car suddenly came to a halt at the corner of a street. A hack running behind, on the track, failed to stop at the same time, and, its impulse being continued, the pole of the vehicle struck D. in the abdomen, near the umbilicus, pressing him with great violence against the back of the car. On admission, the patient was suffering somewhat from shock, and the abdomen was exceedingly tender at the point of injury.

The day following, inflammation set in. The abdomen enlarged, and was so tender that the weight of the bed-clothes could scarcely be borne. Peritonitis, in all its phases, was well marked. Death took place on the fourth day. A *post-mortem* examination showed that a portion of the small intestine was much bruised, but its walls had not been torn through.  Pus and lymph in considerable quantities covered the intestines, gluing them together in several places.

When the liver and kidneys are ruptured, there is usually more collapse than in injury of the intestines. The patient rarely lives long enough to develop peritonitis.

A puncture or rupture of the bladder is succeeded by peritoneal inflammation.  The urine may pass into the abdominal cavity or into the abdominal walls.  In the latter case, the wound is below the part where the peritonæum is reflected over the organ.  If the laceration is at the base, the point of a catheter may pass through and be felt in the rectum.  The escape of urine into the peritoneal cavity is attended with a sharp pain, which rapidly increases till the peritonæum, through its extent, is involved in inflammation. In the cellular tissue of the pelvis or groin, it excites diffuse suppurative inflammation.

*Treatment.*—When the urine accumulates in the cellular tissue, free incisions are made to give it exit.  It is prevented from accumulating in the bladder by allowing it to run out through a catheter introduced for that purpose.  Opium, in full doses, is beneficial.

WOUNDS OF THE PERINÆUM.—Lacerated wounds of this part occur frequently in women during labor.  The child's head, as it is forced down by the uterine contractions, is pressed against the distended perinæum, and, if it is at all resistant, rupture takes place.  As soon as labor has terminated, the edges of the wound should be brought together by sutures.

In the male, these wounds are liable to injure the urethral canal, and operative measures are necessary to relieve the resulting retention of urine and effect a cure. Perineal section is usually performed.

When the patient has been fully anæsthetized, a staff or steel sound is passed down to the laceration and through it, if possible, and the tissue of the perinæum divided in the median line down to that point. The external incision extends from the termination of the scrotum to within half an inch of the anus. The knife is then carried on in the direction of the urethra until the injured portion has been passed. A catheter is then introduced into the bladder and retained for forty-eight hours, to keep the canal open and allow free passage of the urine. A steel sound is afterward occasionally introduced to prevent narrowing of the urethra. As this operation is performed in its most difficult point without a guide, the anatomical relations must be borne in mind. The urethra passes through the triangular ligament from three-quarters to an inch below the pelvis. The opening in this ligament, when appreciated by the touch, will be sufficient to keep the operator from cutting in wrong directions. When a deep, perineal wound bleeds profusely, and the vessels cannot be tied, a small Barnes dilator may be pushed into the opening and filled with ice-water. Dr. Synott, one of the Bellevue house-surgeons, first employed this method. It has proved successful. Another plan is to place a piece of oil-silk, or other suitable material, around a lead-pencil, pass it into the wound, and pack tightly between the oil-silk and pencil a quantity of lint. Ice-bags may afterward be applied to the wound to prevent inflammation. If the blood from the urethra flows out at the meatus urinarius, a sound is passed down the canal and the penis compressed against it with a bandage.

Fractures of the pelvis are sometimes associated with lacerated wounds of the perinæum. The following case is a good illustration.

Patrick C., aged forty; occupation, laborer; was injured while exposing himself in an unnecessary manner over the end of a dock. A ferry-boat, coming into the slip at the time, crushed him against the timbers of the wharf. He was brought to Ward 16, Bellevue Hospital, a few hours afterward. An external examination failed to detect a fracture. A catheter was introduced, but met with an obstruction about the termination of the membranous portion of the urethra. As there was considerable urine in the bladder, it was decided to perform perineal section without delay. Ether was administered to the patient. An incision was then made through the tissues in the median line, commencing near the base of the scrotum and carried within half an inch of the anus. When I reached the membranous portion of the urethra, I found fragments of bone pressing upon, and completely obliterating, the canal. The ramus of the pelvis, and a portion of the body of that bone, were broken in several fragments. The *débris* of soft tissue and bone blocked up the rest of the urethra to the bladder. An opening was, however, made into the organ, and the obstruction removed. The amount of fracture and destruction of tissue rendered his case hopeless. Inflammation set in afterward, and the patient died on the third day.

PENETRATING WOUNDS OF JOINTS, and non-penetrating contused wounds, are always serious. They may result in synovitis, complete or partial anchylosis, or loss of the whole limb. The joint is known to be perforated by the appearance of a thick, transparent fluid (*synovia*) from the joint. This may be absent when the wound passes into the part from above downward.

*Treatment.*—If the wound is small, the edges should be drawn together as closely as possible and held in close apposition by adhesive plaster. Ice-bags, applied afterward, may prevent, or at all events modify, the amount of inflammation. Large wounds should not be entirely closed. Inflammation of the joint is an invariable accompaniment, and a space must be left through which the discharges may pass.

· GUNSHOT-WOUNDS.—Under this head are included all wounds which result from the explosion of gunpowder. They may be made with bullets, cannon-balls, or splinters of wood and stone. The worst wounds are those inflicted by cannon-projectiles and splinters.

All gunshot-wounds, whether external or internal, are attended with danger. A greater amount of shock, contusion, and laceration, accompanies gunshot-wounds than is found in other varieties. Inflammation and suppuration follow in the track of the bullet. Pus is liable to be retained and burrow in the neighboring tissues. Deep suppuration is one of the principal dangers. The wound made by the bullet is smaller where it enters than where it leaves the body, and its edges are inverted, while at the point of exit the edges of the wound are everted. A bullet is easily driven out of its course by bony projections. The missile may strike a rib on the left side, and, passing under the tissues, emerge on the right side of the body. Henner relates a case where the bullet entered the upper portion of the arm and passed down to the thigh on the opposite side.

*Treatment.*—The first efforts of the surgeon are directed to control the hæmorrhage, and to arouse the patient from the state of collapse by stimulants. When this is done, foreign bodies, such as pieces of clothing, bullets, splinters of wood or bone, are to be extracted. The presence of a bullet may be made out in deep wounds by the use of Nélaton's probe. This instrument consists of a silver shaft and a bulbous extremity formed of porcelain. When the bullet is touched a leaden-colored mark is produced on the porcelain. The wound is afterward syringed with a weak solution of carbolic acid, and covered with cloths dipped in an ice-water solution of the acid. Ice-bags are then found serviceable in limiting the amount of inflammation. When suppuration commences, warm fomentations may be used to hasten its progress, and the *débris* prevented from remaining by frequent syringing. In the suppurative stage, there is great danger from secondary hæmorrhage. Therefore, when the wound is in the vicinity of large vessels, it should be carefully watched, and a compress or tourniquet should be placed loosely around the limb, ready to be used at a moment's warning.

Gunshot-wounds of viscera are treated in the same manner that ordinary wounds are after extraction of foreign bodies.

# CHAPTER V.

*WOUNDS OF ARTERIES AND VEINS.*

Ligation of large Arteries.—Air in Veins, etc.—Causes of Sudden Death.—
Treatment.

WHEN large vessels are wounded, there is a great and immediate danger to life. The blood may be poured out externally, or become diffused in the tissues near the artery, or dissect up the sheath of the vessel. Efforts should in every case be made to tie both ends of the bleeding artery in the wound. (*See* article on Hæmorrhage.) If this cannot be done, the artery is then tied between the wound and the heart. Ligature of large vessels is generally followed by complete obliteration of their canals. The ligature divides the middle and internal coats, and brings the external walls together. The blood coagulates at each end of the ligature. The coloring matter of the clot is absorbed. Lymph is poured out between each coat of the artery, between the clot and the lining membrane, and external to the vessels, blending all these parts together, and becoming ultimately a fibrous cord. The ligature, meanwhile, makes its way out by a process of ulceration, and the space formerly occupied by it is filled up by granulation. From ten to fourteen days after the operation the ligature comes away, and then there is the greatest danger of secondary hæmorrhage.

5

As wounds may involve any of the arteries in the body, a short description of the operation in different locations, upon important arteries, will be necessary in this connection.

In wounds of the common carotid or subclavian, it may be necessary to place a ligature on the arteria innominata, an operation rarely attended with success.

When the patient is fully anæsthetized and in position, an incision about two inches in length is made along the inner edge of the sterno-mastoid muscle to the articulation of the clavicle with the sternum, meeting it with a second incision commencing about half an inch from the posterior border of the same muscle, and carrying it along the clavicle. When the integument is turned back, the platysma myoides and sterno-mastoid muscles are divided on a director, the platysma being first cut. The handle of the scalpel is now used to push aside some thick cellular tissue, and the sterno-thyroid and thyro-hyoid muscles are brought into view and carefully divided. A plexus of veins, composed principally of branches of the inferior thyroid, next appears, and must be moved upward and kept out of the way. A thick layer of deep cervical fascia is next incised; the fingers can now be carried down, using the common carotid as a guide, until the arteria innominata is reached. This vessel is situated behind the right sterno-clavicular articulation of the right side. The right vena innominata, internal jugular vein, and pneumogastric nerve, are displaced to the right, and the left vena innominata pressed downward and to the left. An aneurism-needle, armed with a ligature, is then passed around the vessel from below upward.

The *common carotid* artery is ligated either above or

below the omo-hyoid muscle. When the vessel is ligated above the omo-hyoid, an incision is made from the angle of the jaw to the cricoid cartilage. This incision is carried three inches farther than this point when the artery is tied below that muscle. The inner edge of the sterno-mastoid is the guide for both incisions. The integument, superficial fascia, platysma, and deep fascia, are cut through (the three latter on a director); the descendens-noni nerve is moved aside, and the sheath of the vessels lifted with a forceps and opened. The internal jugular vein swells up in the wound as the sheath is cut; it should be compressed above and below the opening, and drawn outward. The pneumogastric nerve is situated here between the artery and vein, and on a plane posterior to both, and great care is necessary to avoid it in passing the ligature. The needle is carried from without inward around the artery. In ligating the carotid on the left side in its lower portion, the jugular vein will be found to have altered its relation to the artery. Instead of lying external to it, it crosses in front of it. Another point to be remembered in connection with the operation below the omo-hyoid is, that the sterno-mastoid artery and the middle thyroid vein run along in the course of the incision, and must be avoided. The sterno-thyroid and sterno-hyoid are drawn toward the median line of the neck. Ligation of the common carotid arteries is sometimes followed by hemiplegia.

The *subclavian artery* is usually ligated in the third portion. In this operation the shoulder is depressed as much as possible, the integument drawn down on the clavicle, and an incision made through it, extending from the anterior margin of the trapezius to the posterior border of the sterno-mastoid. The fascia and platysma having been divided, the external jugular vein is seen near the edge of the sterno-mastoid muscle, and the supra-scapular and transversalis colli nerves and vessels running across the space. These are pushed aside, the deep fascia scratched through, and the finger of the operator carried along the edge of the scalenus-anticus muscle to the tubercle of the first rib, at which point the subclavian artery will be found. The aneurism-needle is carried around the vessel from before backward, and the ligature tied.

The third portion of the *axillary artery* is the most convenient part for ligation. An incision is made about two inches in length, over the head of the humerus, near the centre of the axillary space. The integument and fascia are cut through, the axillary vein drawn inward, the median nerve outward, and the ligature passed from within outward.

The *brachial artery*, in the upper part of its course, is exposed by cutting through the integument and fascia at the inner margin of the coraco-brachialis muscle. The ulnar and internal cutaneous nerves, which lie at the inner side of the artery, and the median nerve, which is situated externally, are separated from the vessel, and the ligature applied.

The brachial may also be tied at the bend of the elbow. The incision is made at the inner border of the biceps muscle. At this joint the artery lies internal to the tendon, with the median nerve still farther inside, close to the vessel. The median basilic vein passes over the artery, separated from it by the bicipital fascia.

The *radial artery* should not be tied at its upper portion, because of its depth from the surface. In the middle third it is exposed by cutting along the inner margin of the supinator longus. The radial nerve, a continuation of the muscle spiral, is found in close relation with it externally. The ligature is passed from the radial to the ulnar side.

In the lower portion of the forearm, the artery is found between the flexor carpi radialis and supinator longus. It is superficial at this point, and easily tied by cutting between those two muscles.

The *ulnar artery*, in its lower portion, is located between the flexor carpi ulnaris and the flexor sublimis digitorum. The ulnar nerve is found at the inner side of the former muscle. The incision is carried through the integument and fascia between these muscles, and the artery tied.

WOUNDS OF THE PALMAR ARCH are difficult to manage, owing to the numerous anastomoses of the arteries. The hæmorrhage may persist after ligation of the ulnar, radial, and brachial arteries. Some surgeons keep a compress on the wound for two or three days, and, if this does not succeed, ligate the vessels in the forearm or arm. When compression fails, the bleeding vessels should be *tied in the wound*, if possible.

*Ligation of the femoral artery* is commonly performed in the lower portion of "Scarpa's space." The integument and fascia are divided at the inner margin of the Sartorius muscle. After the sheath is opened, the femoral vein will be found at the inner side of the artery. The ligature is carried around from within outward.

After ligation of the femoral artery, the limb should be encased in a thick roll of cotton, to keep up its normal temperature, until the collateral circulation is established.

*Ligation of the popliteal artery.*—This vessel is rarely tied except for wounds which involve its walls. In the upper third of the artery the operation is performed by cutting the integument and fascia, at the edge of the semi-membranous. The muscle is drawn inward and the artery exposed. The popliteal vein is external, and superficial to the artery, and the internal popliteal nerve external and superficial to the vein.

In the lower third, the incision is made in the median line, immediately behind the joint. The deep fascia is here very thick, and there is considerable cellular tissue around the vessels, which requires some time and trouble to clear away, so as to bring them into view. When this has been done the limb is flexed, and the needle passed around the artery from without inward.

The *anterior tibial artery* is usually tied in its lower portion above the ankle-joint. The artery is here found between the tibialis anticus and extensor proprius pollicis, and is covered by the integument and fascia. These latter are incised—the tendons separated, and the artery exposed. The nerve is in this situation superficial to the artery. The venæ comites are separated from each side of the vessel, and the ligature applied in the usual manner.

*Posterior tibial.*—It is extremely difficult to reach this artery in its middle third, because of its depth from the surface. The operation is performed by extending the foot, making an incision at the inner border of the tibia about three inches in length. When the integument and fascia have been cut, the edge of the gastrocnemius muscle is turned aside, and the soleus detached from the tibia by cutting its fibres on a director. The fascia underneath this muscle is next divided, and the artery exposed from three-quarters of an inch to an inch from the inner border of the tibia.

The tibial nerve in this region is situated on the outside of the artery, and should be separated from the vessel before tying.

The vessel is sometimes tied as it passes around the ankle, by making a curved incision midway between the internal malleolus and the heel. The integument and superficial fascia having been divided, the needle is passed from without inward, as in the previous case.

WOUNDS OF VEINS, ENTRANCE OF AIR.—Fatal hæmorrhage takes place in a short time when large veins, as the jugular or vena innominata, are wounded, unless immediate assistance is rendered, and the wound closed by ligation or pressure. In wounds of small veins the danger from hæmorrhage is slight.

Wounds of veins may be followed by phlebitis or by the entrance of air. The latter complication occurs particularly in the veins of the upper extremity and neck, during operations for the removal of tumors. The air enters the opening in the vein with a loud hiss, and the patient, in many cases, expires instantly. If only a small quantity of air enter, there is a tendency to syncope, difficult breathing, and convulsive movements of the body, which may last for several hours before a fatal termination is produced. In the majority of cases sudden death ensues.

A number of explanations have been offered to account for the suddenness of death in this accident. Bell thought it due to the action of air upon the medulla oblongata. Moore ascribed it to irregular action of the valves of the heart from the presence of air; * others, again, ascribed it to the impossibility of a frothy liquid passing through the lungs.

In the absence of any accepted theory, I would suggest the following : In the great majority of cases the accident occurs in removing tumors from the neck or axillary region. These tumors by their pressure empty the veins upon which they lie. As the knife of the surgeon passes into the vein, and the weight of the tumor is removed, air rushes in to fill up the vacuum, and the heart ceases. When it is considered that the pressure of the atmosphere is equal to fifteen pounds to the square inch, and the force-pump action of the heart only thirteen pounds and a half to the square inch, it will be seen that the column of air by its own direct pressure is sufficient to overcome and paralyze the muscular force of the heart. The stoppage is instantaneous. Subsequent pressure on the wound fails to do good, because of the presence of air in the heart, which cannot be disposed of with sufficient rapidity to enable the organ to recover itself. The distention of the right side of the heart, which is usually found after death, is accounted for on these grounds.

When only a small portion of air enters, and pressure is made on the wounded vein, there is sometimes recovery.

Whenever operations are performed about the neck or axilla, every vein in the vicinity of the surgeon's knife should be closed by assistants. Both before and after the removal of the tumor, this precautionary measure is called for.

*Treatment.*—Immediate efforts to restore the respiratory movements, and with them the action of the heart, should

* Holmes's Surgery, article Wounds of Veins.

be made. Marshall Hall's or Sylvester's methods of artificial respiration can be tried. Stimulant enemata and
friction of the surface are always necessary. Galvanism
may also be tried. In mild cases, brandy and ammonia
may be given by the stomach. Hot plates over the epigastric and precordial regions are also serviceable.

# CHAPTER VI.

Dissecting Wounds.—Hydrophobia.—Snake-Bites.—Insect-Bites.

Dissecting Wounds.—During the process of putrefaction a poison is generated which is capable of exciting inflammation in healthy tissues, and of reproducing itself in the circulation, giving rise to serious constitutional disturbances. The poison is introduced by cutting or puncturing the flesh with the knife used during the progress of *post-mortem* examinations, or in the anatomical investigations of the dissecting-room. Wounds of the most serious character may be made by a piece of broken rib or other rough bone.

When putrefaction is much advanced, the system is less likely to be infected. It is an established fact that wounds inflicted in the dissecting-room, when decomposition is nearly at its maximum, are comparatively harmless, while those inflicted in a *post-mortem* examination often destroy life. Whether the material injected in the arteries of subjects about to be dissected modifies the poison or not, is a subject for future investigation.

The disease with which the patient died has much to do with the severity of the disease in the wounded person. Puerperal fever, erysipelas, pyæmia, typhus, etc., are peculiarly dangerous. They seldom fail to produce either local or constitutional poisoning. On the other hand, parturient women are sometimes infected by the poison of the dissecting-room carried on the hands of a physician. Erysipelas, puerperal fever, etc., are not unfrequently developed in this manner.

Debilitated states of the system are favorable to the infection. The influence of the poison is more strongly manifested in every case where the constitution is below par.

In merely local poisoning, the wound shows little tendency to heal, closing for a day or two and then breaking out afresh. Around the wound the integument is thickened, and of a dusky hue. There is an exudation from the cut surface, of a sero-purulent character. This condition of the wound may last for weeks, and even months, healing partially for a time, then breaking out and assuming its original unhealthy appearance.

In another variety the wound, after a lapse of twenty-four or thirty-six hours, becomes hot and painful. A small quantity of sanious fetid pus exudes from the surface. The surrounding integument is red and swollen. In a short time, small red lines may be noticed running up the arm, indicating the extension of inflammatory action to the lymphatic vessels (angeioleucitis). The arm is swollen and painful. The axillary glands enlarge and often suppurate. Abscesses may form and burrow in the cellular tissues of the arm and chest. The skin is hot and dry, the pulse rapid, and urine scanty and high-colored. When the abscesses open and discharge, great prostration ensues, which may destroy the life of the patient or leave him a helpless invalid for months.

The third class of cases rarely recover. The patient,

within a period ranging from twenty-four to forty-eight hours after the wound is received, is seized with violent chills. These are succeeded by unmistakable evidences of blood-poisoning. The pulse becomes rapid and very small, the countenance anxious, and tongue brown and dry. The integument is of a tawny color, and may be jaundiced. There is profuse perspiration. Meanwhile, the wound becomes very painful; the tissues around it are thickened and infiltrated with pus. Abscesses are not confined to the injured tissue, but may show themselves in any part. The lymphatics are involved as in the preceding case. Delirium sets in, and is soon followed by death. In severe cases, death may occur within forty-eight hours after the infliction of the injury.

*Treatment.*—In wounds of this character, proper precautions should be immediately resorted to in order to prevent the retention of the poison and its subsequent entrance into the circulation. The wound should be washed by holding it under a stream of water for a few seconds. The lips are then applied and the virus removed by suction. There is no necessity for the application of caustics.

The treatment of cases where there is only local poisoning resolves itself into stimulation of the wound by means of carbolic-acid or nitrate-of-silver solutions, and maintaining the health of the patient at a proper standard, by fresh air, good food, and tonic medicines.

In those cases where acute inflammation appears in the wound and extends to neighboring tissues, the wound should be enlarged and cleansed of accumulations of pus with a strong solution of carbolic acid. A poultice of linseed-meal and charcoal may be then applied to the wound, and, if necessary, to the whole limb. Painting the inflamed lymphatic vessels with iodine has been recommended.

Opium is freely given to relieve pain and to produce sleep. Easily-digested nutriment, such as beef-tea and chicken-broth, is to be administered *ad libitum*. Stimulants are sometimes necessary. The treatment for the third variety is similar, with the addition of stimulants used freely, and large doses of quinine.

HYDROPHOBIA.—Phobodipson, rabies, canine madness, lyssa, and a variety of other terms, have been used to designate this malady. It has been known from the earliest historical periods. The disease attacks man and many of the lower animals. Dogs, cats, and wolves, are most subject to its ravages. Cows, goats, pigs, and horses, are occasionally afflicted. It occurs at all seasons of the year, without reference to climate or temperature, appearing in the winter season as well as in "dog-days." The nature of the poison is unknown. It is transmitted from one animal to another by means of the salivary secretions introduced through wounds inflicted by the teeth. Other secretions in the body are said to be harmless and unable to transmit the disease.

The period between the inoculation and the development of the disease is subject to considerable variation. Generally it appears between one and two months. Cases have been recorded (hardly with sufficient authority, however, to establish them as facts) where the disease remained latent for twelve or fifteen months.

Billroth mentions an old superstition which attaches great importance to the number nine, and gives the disease a tendency to develop on the ninth day, ninth week, or ninth month, succeeding the injury.

Rabies in the dog is divided by Virchow into three stages: 1. The melancholic; 2. Furious; and 3. Paralytic. The animal affected loses its appetite—shrinks from water and ordinary food—endeavors to hide in his kennel, and can with great difficulty be coaxed out. The head droops, and the eyes are bloodshot and heavy. There is great thirst, and water is not refused.

In the second stage the animal yelps or howls, and runs wildly about, biting at every thing. The tongue hangs from the mouth, and the eyes are congested and wild.

In the third stage emaciation is apparent and rapidly progresses, great exhaustion supervenes. Little effort is made to move, and the saliva dribbles from the mouth. In walking, both hind-legs are dragged on the ground as if paralyzed. Death ensues in from four to eight days from the commencement of the disease.

HYDROPHOBIA in man has many of the characteristics just described.

A person bitten by a mad dog is usually on the watch for some manifestation of the disease. The wound may heal readily, but the dread remains. If the cicatrix begins to inflame and is painful, and other signs appear which show that his fears are about to be realized, the depression of spirits and anguish are intensified. All cases are preceded and accompanied by this terror. It is one of the characteristics of hydrophobia.

As the disease progresses, the skin becomes hot and dry, the pulse rapid, and lacking strength. There is much thirst. In two or three days from the first manifestation of the disease the muscles of the throat, and especially those concerned in deglutition, become stiff and sore. Attempts at swallowing are followed by spasmodic contraction of these muscles, and of those concerned in respiration. These convulsive movements increase in frequency, excited by the smallest provocation. Slamming doors, cold currents of air, pouring water from one vessel to another, or changing the bedclothes, brings them on. In some cases there are general convulsions. Thirst is intense, and the unfortunate patient does not relieve it for fear of choking or renewing the spasms. Sometimes there are small pustules under the tongue (Marschetti). The patient's countenance expresses all his terror. The eyes are staring and bloodshot. A thick saliva is constantly thrown from the mouth. The voice is husky. As the end approaches, the skin becomes cold and clammy, the pulse almost imperceptible, and the respiratory movements irregular. A convulsion may terminate life by involving the muscles of respiration, or the patient may die gradually from exhaustion.

After death, the fauces, throat, and lungs, are dark-colored and congested. In some cases, there are congestion of the cord and effusion into the ventricles of the brain. There is nothing definite in any of the lesions to indicate the specific action of the virus.

Strange as it may seem, hydrophobia is sometimes imitated for mercenary purposes. A case of this kind was admitted to Ward 9, Bellevue Hospital, in the winter of 1867. The patient stated that, when seven years of age (he was then twenty-five), he was bitten by a mad dog. One year afterward, symptoms of hydrophobia manifested themselves. He recovered from that attack, but exactly one month afterward at "the full of the moon," he was affected in a similar manner. This peculiar tendency to a monthly recurrence kept up for two or three years, and then ceased up to within two years of his first appearance. At that time they again commenced, and had continued at irregular intervals until his admission to the hospital.

While in the reception-room, awaiting transference to the ward, an orderly approached him with some water, which immediately threw him into a convulsion. He writhed violently on the floor, throwing the arms and legs about in every direction. The saliva collected in the form of foam around his mouth, and he howled and yelped like a "mad dog." The convulsion lasted for two minutes. At its termination he seemed to be quite exhausted, but was able to walk to the ward.

Shortly after his admission, and while in a convulsion, he was seen by Dr. Flint, who advised the application of hot water to the skin. The patient did not wait for the remedy, but recovered immediately. Finally, after a close questioning, he confessed the fraud, and admitted that for many years he had practised the game successfully, making considerable capital out of it.

This man's story was told with such an appearance of candor, that it was hard to doubt at least his own faith in the reality of the disease.

*Treatment.*—A wound inflicted by a dog suspected of madness should be washed and sucked as in ordinary dissecting wounds, and afterward thoroughly cauterized Complete excision of the part is better, in most cases, than destroying the tissues by cauterization. Previous to the washing and excision, some recommend that a ligature be placed tightly around the limb, above the wound, in order to prevent absorption of the poison. On the arm or leg the procedure is useless, because the circulation through the deep veins cannot be completely stopped. If placed on the fingers or toes, it may answer. In the bitten parts the excision should extend some distance into the healthy tissue, and the wound be subsequently cauterized. The actual cautery is the best, but the most painful.

When the disease is fully developed but little can be accomplished. Stimulants can be given in large quantities by enema, and other liquids in like manner. Opiates and anæsthetics should always be administered to relieve the pain and distress, and decrease the convulsive movements. As the wound has again become inflamed and painful, hot disinfecting poultices, sprinkled with laudanum, will be serviceable. Free discharge should be kept up continually.

SNAKE-BITES.—Among the principal venomous reptiles may be enumerated the whip-cord snake, cobra de capello, rattlesnake, viper, and adder. The bites of the first two produce a fatal result more quickly than the others. Rattlesnake-bites stand next in order of virulence. Viper and adder bites are fatal only to very young animals, or to children of tender years. In the more deadly classes the symptoms following a bite, and the action of the poison, are the same.

Rattlesnake-bites are not uncommon in the Southern and Western States, and the mortality attending them is very great.

The venom of this reptile is contained in a small sac situated at the base of the sharp tooth or fang. The tooth is channelled throughout its centre to make a place of exit for the poison. When the tooth is inserted into the tissues, the

6

poison-sac is compressed, and the venom ejected into the wound.

The person bitten is overcome, either immediately or after the lapse of a few minutes, by a feeling of faintness and great depression. The pulse becomes feeble, rapid, and intermittent. The pupils are dilated ; there is some pain over the abdomen, vomiting, and sometimes purging. Delirium is present in most cases. The extremities and surface of the body are cold and clammy, respiration is catching and difficult. Coma comes on, grows rapidly deeper, and terminates in death.

The wound, shortly after the bite, swells rapidly. In one case it assumes a dark-red color, in another a bluish-black. A few patches of a light color may be intermixed. There is a sharp, intense pain in the wound, which extends up the limb, generally in the course of the principal nerves. Inflammation extends to the neighboring tissues, and, if the patient live long enough, diffuse suppuration may occur, and abscesses form throughout the limb.

Rattlesnake-bites produce death in from five to ten hours. The *post-mortem* appearances show nothing of the special effects of the poison. Sometimes there is congestion of the brain, with serous effusion underneath the arachnoid and into the ventricles. There may also be congestion of the lungs and mucous membrane of the stomach and intestines. The blood remains fluid in the cavities of the heart in many cases.

*Treatment.*—The wound should be treated in precisely the same manner as a wound produced by the bite of a mad dog; that is, the part should be washed, sucked, excised, or cauterized.

A vast number of internal remedies have been proposed. Bilron's antidote is one which has been strenuously advocated. Dr. W. A. Hammond, after a series of experiments, came to the conclusion that it was a remedy of great efficacy. Its formula is as follows :

℞. Potassii iodidi . . . . . . . gr. iv.
  Hydrg. bichloridi . . . . . . gr. ij.
  Bromii . . . . . . . . ℥iv.

From ten to twenty drops of this mixture are given every half-hour, until an amelioration of the symptoms is produced.

Arsenic is another remedy highly spoken of. Guaco, Virginia snakeroot, and other medicines of vegetable origin, have also acquired temporary reputation as antidotes. The the most efficacious treatment is to administer large doses of carbonate of ammonia repeatedly in conjunction with enemata of whiskey or brandy. The ammonia can be administered in ten or twenty grain doses every half-hour. Friction to the surface, with hot pieces of flannel dipped in alcohol, is also beneficial.

The poisoned wounds produced by scorpions, tarantulas, centipedes, and other members of this class, are rarely attended with destruction of life.

*Scorpions* have an elongated body and a slender tail, the latter six-jointed. In the last joint there is a sharp sting, which communicates with poison follicles. Scorpions are found in all tropical climates. The largest scorpions are the most venomous.

The tarantula, a species of spider which inhabits Southern Europe, was at one time held in great terror on account of its reputed deadly influence. The stories of its ravages are, however, not founded on fact.

Centipedes are less dangerous than either of the preceding varieties. The most venomous grow to a length of six inches. A number of poison-claws project from the body. As the insect crawls over the surface, these are inserted into the integument, and the virus introduced. Some writers deny the existence of any special poison in members of this class.

The constitutional symptoms following the bites of these insects are exhibited in the form of headache, vertigo, dimness of vision, and sometimes febrile excitement. The wound, in some cases, is not inflamed; in others, it becomes red and painful, and the inflammation spreads to other parts of the extremity injured, ending in diffuse suppuration.

*Treatment.*—When the wound is cleansed, it should be sponged thoroughly with a strong solution of ammonia, and afterward covered with cloths moistened with the same substance. Brandy may be given internally in conjunction with ammonia.

# CHAPTER VII.

Foreign Bodies in the Larynx, Trachea, Bronchial Tubes, Pharynx, Œsophagus, Eyes, Nose, Ears, Urethra, Bladder, and Rectum. — Tracheotomy. —
Laryngotomy Laryngotomy.—Œsophagotomy.

FOREIGN BODIES IN THE AIR-PASSAGES.—Foreign bodies
are usually lodged in that portion of the air-passages known
as the larynx. This organ is situated in the median line of
the neck, between the trachea and base of the tongue. The
anterior margin of its superior opening is guarded by a cartilage called the *epiglottis.* During the act of deglutition,
the epiglottis closes the aperture in the larynx, and prevents
the entrance of food as it passes over on its way to the
œsophagus. It is raised during the respiratory movements
for the free ingress and egress of air.

The trachea commences opposite the fifth cervical, and
bifurcates about the third dorsal vertebra into the right and
left bronchus. The right bronchus is shorter than the left.
Its orifice lies directly under the tracheal canal, so that foreign bodies which pass below the trachea drop in and effect
a lodgment. The endeavor to talk, laugh, or respire, with
food or other substances in the mouth, is often followed by
the entrance of some portion into the air-passages. In talk-
ing or laughing, the air is passing out of the lungs, and the
epiglottis is raised. Heavy substances contained in the

mouth during these acts, readily roll backward, notwith-standing the outward current. Taking a sudden inspiration while eating is more dangerous, as the current of air pass-ing downward is liable to sweep a portion of the food along with it. Vomiting, while in a state of intoxication, is apt to be attended with the entrance of half-digested particles of food into the larynx. It is not unusual for worms to find their way into the larynx during sleep, or for bronchial glands to become detached and carried upward, producing serious and even fatal results. The presence of a foreign body in the pharynx, or œsophagus, may induce spasm of the glottis, and lead to the erroneous supposition that it has found lodgment in the air-passage. The introduction of a probang will settle the difficulty.

Children are more often subjected to this accident than adults are. The habit of carrying in the mouth beads, marbles, or pennies, is very prevalent among them. As an instance of the dangerous results attending it, the following incident, which occurred in Bellevue Hospital, may be of interest:

While engaged in amputating the great-toe of a little girl, who was under the influence of chloroform, she sud-denly ceased to breathe; the face assumed a purple hue, and death seemed imminent. Apprehending that the chloro-form was the cause of the difficulty, I commenced artificial respiration. While I compressed the chest, my assistant introduced his finger into the mouth to clear the throat of mucus, and draw forward the tongue. In so doing he found a copper coin completely closing the superior aperture of the larynx. The removal was soon followed by a renewal of the respiratory movements, and disappearance of all the alarming symptoms. The child had been playing with the penny, and had placed it in her mouth previous to my arrival in the ward, and, when insensibility was induced by the anæsthetic, it fell back into the larynx.

Foreign bodies may lodge in the upper part of the larynx—in the ventricle between the vocal cords, or in the trachea and bronchial tubes. The symptoms differ with the location of the material, and the length of time it has remained.

The size of the foreign body bears no special relation to the severity of the symptoms, unless, indeed, it is so large as to completely block up the canal. A light substance capable of being moved up and down with the respiratory movements occasions greater distress than one which is stationary. When the material lodges in the larynx, whether large or small, it produces a spasm of the laryngeal muscles which close the glottis, and thus prevents the passage of air. The patient struggles for breath, the lips and cheeks become livid and swollen, the eyes protrude from their sockets, convulsive movements of the limbs accompany the agonizing efforts to breathe, and the patient dies at once, or receives temporary relief from a relaxation of the spasms. The current of air which now enters, either passes the obstruction, or carries it farther down into the trachea. Once in this organ, the intense suffocative symptoms become less marked and continuous. There is more or less difficulty of respiration all the time, pain over the point where the foreign body is lodged, and a distressing cough. The countenance has an extremely anxious expression; the pulse is rapid. Severe dyspnœa occurs now only at intervals. Whenever the substance is forced up into the larynx, violent efforts at

expulsion again ensue, with the same paroxysm as characterized the first stage.

When the foreign body reaches one of the bronchi, the
lung on the corresponding side gives but little respiratory
murmur on auscultation, and over the opposite lung there
are exaggerated respiration and increased resonance on percussion.

The presence of a foreign body in any part of the air-
passages gives rise to symptoms like those mentioned above
—they only differ in degree.   After a day or two has
elapsèd we have more pain—the cough is increased, the
pulse becomes accelerated, the countenance retains its anxious expression, the voice is husky, and general febrile action is developed.   There are also the special signs of inflammation in the part occupied by the irritating material.

Death may occur instantaneously in the *first period*,
from asphyxia or injury to the brain, from extravasation of
blood following the violent efforts to respire.   In the second
period death is induced by bronchitis or laryngitis.   If
weeks and months elapse before its expulsion, abscesses may
form, and the patient succumbs to exhaustion.

*Treatment.*—A violent blow on the back, if given immediately after the accident occurs, will assist the natural
efforts of the patient in ejecting the foreign body.   After it
has passed the larynx, this procedure alone will be of little
avail.   If the first attempt fails, the body is to be inverted
and held up by assistants, while the physician strikes with
the open hand between the shoulders, at the same time
moving the patient rapidly from side to side.   If this
method induces violent suffocatve paroxysms, it must not
be repeated.   Should the urgent symptoms continue, which

they are liable to do, laryngotomy or tracheotomy must be performed without delay. The acute sensibility of the larynx hinders the irritating material from passing the glottis, which closes spasmodically every time it reaches that point, and, unless an opening is made lower down to give it exit, death may soon ensue. Some surgeons advise the administration of emetics, but such practice is worse than useless.

Tracheotomy is preferred above other operations by some practitioners, especially for children; but, if circumstances admit, laryngotomy should be first performed. It possesses many advantages worthy of attention :

1. The parts are more accessible at all periods of life.

2. It is performed with greater rapidity, and consequently is peculiarly applicable to cases requiring instant relief.

3. There is no danger of wounding important vessels, or delaying the operation by hæmorrhage.

*Laryngotomy* is performed through the membranous interval existing between the thyroid and curved cartilages. The region is superficial and readily exposed. The only vessel to be avoided is the crico-thyroid artery, which passes across the upper part of the space to anastomose with its fellow on the opposite side.

The patient should be placed in a chair or in the recumbent posture, with the head thrown back, and the larynx steadied by an assistant. An incision about an inch in length is made through the integument over the crico-thyroid space, fully exposing the membrane, which is then opened by a transverse cut near the cricoid cartilage. By keeping close to this cartilage, all danger of wounding the artery is avoided. The aperture thus made in the larynx is now widened by a dilator or ordinary forceps, and the patient turned on his chest. If the opening be too small, the incision may be carried down through the cricoid cartilage and upper ring of the trachea.

The ejection of the foreign body often occurs as soon as the operation is completed, but, if this desirable result do not follow, and the substance be within reach, a long-curved forceps may be carefully introduced to remove the obstruction. When the passages are entirely cleared, the edges of the wound must be approximated and allowed to heal.

*Tracheotomy* requires greater care and skill in its performance than laryngotomy. The trachea, especially in children, is deeply seated, and covered by important plexuses of veins and close proximity to large arteries. The parts to be avoided in the operation are: 1. The anterior jugular veins. 2. The isthmus of the thyroid gland which lies on the second and third rings of the trachea; and 3. The inferior thyroid veins.

It is always safe to administer chloroform to a child before commencing the operation. It renders material assistance to the surgeon, by relieving spasm and keeping the patient from struggling. Should it be considered advisable to dispense with the anæsthetic, the child's body must be enveloped in a sheet, which will keep the limbs motionless. The head is thrown back in the former case, and the larynx held by an assistant. An incision is made through the integument directly in the median line, beginning a short distance below the cricoid cartilage, and continued down from one and a half to two inches. By keeping exactly in the median line the anterior jugular veins are avoided. These vessels are pushed aside, and the incision carried through the fascia, which covers the sterno-hyoid and sterno-thyroid muscles. These muscles are separated, and the inferior thyroid plexus of veins is reached. The handle of the scalpel is now to be carefully used in getting them out of the way without laceration. A tenaculum is inserted into the trachea to draw it forward. The knife is introduced between the rings, and two or three of them divided from below upward. The cut-ends are held apart by ligature or widened by dilators, and the patient is placed in a supine posture, and, if the obstruction still remains and is within reach, it must be removed with the forceps.

When these operations are performed for other pathological conditions, as laryngeal inflammations, tumors of the larynx, œdema glottidis, croup, etc., a curved tube is introduced through the opening, and allowed to remain until the difficulty which called for the operation is removed.

When the operation is concluded and the tube inserted, the patient must be carefully watched for a day or two, and the tube kept clear of blood and mucus. The old form of trachea-tube necessitated the use of a feather in order to keep it clean ; but the variety now employed has a second tube fitting closely inside the first, which can be removed and cleaned at pleasure without disturbing the patient.

FOREIGN BODIES IN THE PHARYNX AND ŒSOPHAGUS.—The pharynx is that part of the alimentary canal which extends from the base of the skull to the fifth cervical vertebra, where it becomes continuous with the œsophagus. It lies behind the nose, mouth, and larynx, in the order mentioned from above downward. Its widest part is opposite the hyoid bone, its narrowest portion is where it joins the œsophagus. The food passes into it from the mouth, and is carried down into the œsophagus by contraction of the pharyngeal muscles.

The œsophagus commences opposite the cricoid cartilage, to which it is attached by muscular fibres, and terminates in the cardiac extremity of the stomach, on a level with the ninth dorsal vertebra. In the neck it lies behind the trachea. It measures nine inches in length, and is the narrowest portion of the alimentary canal; the most contracted parts are at its origin, and as it passes through the diaphragm to connect with the stomach.

Various foreign bodies have lodged in the œsophagus and pharynx—among the most frequent of which are bulky articles of diet, such as meat, potatoes, beans, apples, etc., and metallic substances, such as pennies, needles, pins, and nails, and even bones, false teeth, India-rubber, and pieces of glass have been found. The symptoms depend in some degree on the location and character of the foreign body. When of large size, it is apt to stop at the lower portion of the pharynx, and by its pressure on the larynx cause spasm of the glottis and consequent suffocative paroxysms. Should it pass below this point, the pressure on the trachea may still obstruct the entrance of air. After the foreign body fully enters the œsophagus, it generally reaches the lower constricted portion at the cardiac orifice before it again lodges. Small bodies, such as pins or needles, pierce the mucous membrane, and cause more pain and irritation than other varieties. If they stop at the lower anterior part of the pharynx, spasmodic closure of the glottis is induced, often to a greater extent than when bodies of a large size press on the same part. Irregular sharp substances in the pharynx or upper end of the œsophagus cause nausea and vomiting.

In the average of cases there are pain at the point of lodgment or over the episternal notch, and difficulty of swallowing. The patient is often extremely nervous, and complains of general distress in the throat.

*Treatment.*—In all cases of simple obstruction of the pharynx or œsophagus, the first endeavor should be to ascertain the character of the material swallowed and its point of lodgment. The first point can be ascertained from the patient or friends; the second by an examination with the finger, elastic bougie, or probang, and by the seat of the pain. The latter symptom, however, is not reliable, for in many instances the pain remains after the foreign body has been swallowed or vomited. The patient's statements, therefore, cannot be implicitly relied on.

In examining the pharynx, an ordinary laryngoscope may be used with advantage. When the tongue is fully depressed, and the light thrown in, the patient should then take a deep inspiration, which will separate the pillars of the fauces, and allow inspection.

If the obstruction is in the pharynx or upper part of the œsophagus, it should be removed if possible. If below the level of the episternal notch, and not too large or sharp, it may be pushed down into the stomach. Particles of food may generally be treated in this manner when below the point named, or when it is difficult to extract them. The use of dilute mineral acids will soften a piece of bone so that it will go down (*Hall*).

Force must not be employed in removing needles, pins, or other sharp articles, for fear of piercing, or lacerating the mucous membrane, and the important neighboring parts. Obstructions in the upper portion of the pharynx may be extracted with the finger, or when farther down with curved forceps adapted to the purpose.

Among the instruments that are used for pushing foreign bodies into the stomach the *probang* is the best. It consists of a thin strip of whalebone with a piece of sponge attached firmly to one end. It is carefully introduced and moved slowly downward, until the foreign body is reached and dislodged. Elastic bougies or catheters are used in the same manner. When needles or pins become impacted in the canal, an elastic catheter having a skein of silk fastened in the eye may be introduced until it passes below the obstruction; it is then drawn up, entangling the needle or pin in the meshes of the silk (*Gray*).*

A very ingenious instrument has recently been employed by surgeons in this city, for the removal of foreign bodies. It consists of a gum catheter, from which the end has been cut, a thin piece of whalebone several inches longer than the catheter, and a number of bristles. The whalebone is made to slide readily up and down inside the catheter. The bristles are attached by an extremity to the end of the whalebone, which protrudes from the catheter; the other is fastened around the open end of the catheter. When the whalebone is pushed out through the catheter as far as possible, the bristles surround the whalebone very closely and compactly. The instrument in this condition is then carried below the obstruction, and the catheter firmly held, while the whalebone is drawn up within it. This causes the bristles to double up in the centre, and protrude

* Article Foreign Bodies, Holmes's Surgery, vol. ii., page 325.

all around in such a manner, that when the instrument is withdrawn it carries the foreign body with it.

When foreign bodies are not removed, they produce ulceration and suppuration of the parts pressed upon, and other organs become involved. If milder methods fail, we must resort to *œsophagotomy*.

The operation should be performed on the side occupied by the foreign body, or, if this cannot be determined, the left side must be selected, because, in the neck, the œsophagus inclines to the left of the median line, and is therefore more easily reached.

After the patient is fully under the influence of an anæsthetic, the shoulders are raised, the head turned to one side, and an incision is made along the inner border of the sterno-mastoid muscle, commencing on a level with the upper border of the thyroid cartilage, and extending down about four inches, cutting through the integument and platysma-myoides muscle. The omo-hyoid muscle is then exposed, and must be either cut or pushed aside. The sheath of the carotid vessels comes next in view, and is drawn outward and retained by an assistant while the thyroid gland and trachea are moved slightly inward. A bougie is now passed down the throat, and protruded below so as to bring the œsophagus fully to view in the wound. An opening is then made, through which the foreign body is extracted.

The patient should be fed daily through a tube for two or three weeks after the operation, in order to give the œsophageal wound time to heal.

FOREIGN BODIES IN THE NOSE.—Children of tender years are particularly liable to this accident. It is of frequent occurrence, but happily there is more inconvenience than danger attending it.

Peas and beans in the nasal cavities are specially troublesome; they enlarge in size by their absorption of moisture, and by an increase of pressure cause greater irritation. Peas and beans have been known to sprout in the nasal cavities after having remained there for several days, giving rise to serious inflammation of the mucous membrane and spongy bones.

*Treatment.*—Having by careful examination determined which nostril the obstruction is in, snuff or other sternutatory may be introduced into the opposite nostril, in order to induce sneezing. This procedure will probably dislodge the foreign body. In place of this, a stream of water, carried into the nostril by means of "Thudicum's nasal douche," may wash out the material. When simple measures like the foregoing are found useless, the forceps must be employed. The long curved forceps used for the extraction of polypi may be tried. The instrument is passed up carefully to the foreign body, closed upon it and drawn down. In all cases care should be taken that the substance is not forced back through the posterior nares into the throat, or that the efforts at extraction are not carried to such a length at one sitting as to fatigue the child, or cause inflammation in the organ.

FOREIGN BODIES IN THE EAR.—The length of the external auditory canal is about one inch and a quarter, and at its inner extremity is the membrani tympani, a delicate membrane which separates the middle from the external ear. Across the middle ear are stretched three small bones connected externally with the membrani tympani, and, through the foramen ovale, on the inner wall with the internal ear.

Foreign bodies in the external ear, in consequence of their close proximity to important and delicate structures, may produce grave and even fatal results. The inflammation usually excited by their pressure may extend to the membrani tympani, destroying it and causing deafness. It may pass on to the middle ear, involving the temporal bone, giving rise to caries and abscess, and may even reach the brain, exciting fatal meningitis or abscess in the middle lobe of the cerebrum. Sometimes efforts at extraction cause permanent deafness by rupturing the tympanum.

Grains of wheat, corn, seeds, and also insects, such as bugs or fleas, have been found in the auditory canal. Insects cause great irritation, but their removal is not attended with difficulty. Accumulations of wax of any great quantity may cause distress.

If the body is large, there is considerable pain and singing in the ear, and more or less deafness is experienced. If it is allowed to remain in the canal, there will be in the course of twenty-four to forty-eight hours a discharge from the meatus, which soon becomes purulent and mixed with blood.

Small substances do not excite inflammation so rapidly, but are often as difficult to extract as large bodies. Insects create an itching in the canal, and a loud rattling or grating noise, excessively annoying to a nervous individual.

*Treatment.*—Insects are removed by closing up the external meatus, or as much of the canal as possible, and preventing the admission of air. This is best done with a

7

piece of " cotton-wool," thoroughly saturated with a strong
solution of common salt or vinegar, and sufficiently large to
plug the orifice completely.   After its introduction turn the
patient on the affected side, and allow the hand to press
firmly on the ear.   In a few minutes the noise and irritation will cease, and, if the plug at this time be withdrawn,
the insect will probably be found partially embedded in its
substance.

To remove small bodies, a stream of water may be thrown
gently into the canal, or a scoop and bent probe may be
used.   The scoop should be introduced into the *upper* part
of the canal, so that, in pressing on the foreign body, the
edge of the instrument will recede, instead of pressing
against the membrani tympani, as it undoubtedly would if
inserted below.   Great care must be observed in the employment of these instruments, and very little force should be
exerted through them.

If it is found impossible to remove the obstruction by
these means, the canal must be syringed gently twice each
day with warm water, until all inflammatory symptoms
have subsided.   In the majority of cases the foreign body
will come away in the purulent discharge.

FOREIGN BODIES AROUND THE EYE.—Sand, broken eye-
lashes, cinders, etc., often lodge under one of the lids, usually
the upper lid.   If these substances remain, inflammation of
the *conjunctiva* will be established, and ulceration set up
around them.

*Treatment.*—Hairs which have become fixed in the conjunctiva should be extracted with forceps.   To do this, the
lid is everted, and the eye cleansed of any effusion which
may have collected around the hair ; the latter is then readily

removed. For the extraction of dirt, sand, etc., the following simple proceeding will answer : Grasp the upper lid
between the thumb and forefinger, lift it from the eyeball
and draw it forcibly down, outside of the lower lid. When
stretched as far as possible, allow it to slide slowly back to
its natural position, touching its fellow as it goes up, then
wipe the edges with a handkerchief so as to remove the
foreign body from the lashes. The operation can be repeated
three or four times, or oftener, without injury. Some use a
small scoop made from wire, which is moved around under
the eyelid from one canthus to the other.

FOREIGN BODIES IN THE URETHRA AND BLADDER. — In
many cases this occurrence depends on unnatural or uncon-
trolled desires which seek relief in local irritation and excitement. The most astounding means are resorted to for this
purpose. Slate-pencils, hair-pins, knitting-needles, wire,
pieces of wood, leather strips, straw, tobacco-pipes, etc., are
among the long list of articles which have been extracted
from these organs.

Prof. James R. Wood has in his collection a thick leather
thong, with a large knot at its extremity, which a patient
of his was in the habit of introducing into the urethra.
On one occasion the knot passed beyond the sphincter
muscle, and was forcibly held. It had to be removed by
an operation.

However, there are other means by which foreign bodies
become lodged in the urethra and bladder. In the dilatation of a stricture with elastic bougies, or while using a
catheter, the instrument may break, and the pieces remain
impacted.

After remaining a certain length of time in the bladder,

foreign bodies become encrusted with various salts, and grow larger by deposit. Such an occurrence is attended with all the symptoms and dangers of stone. In the urethra they may cause inflammation and sloughing of the mucous membrane, and subsequent stricture.

*Treatment.*—Extraction is necessary in all cases. When impacted in the male urethra, the removal may be effected by a forceps adapted to the canal. If this fail, urethrotomy must be performed. Foreign bodies in the male bladder are sometimes broken up with a lithotrite; but in most cases perineal section (*see* page 61), or some of the operations for stone, are usually made. Substances may be taken from the female bladder with a forceps. The urethra in females is very short and easily dilated, so that the introduction of a forceps or other instrument is accomplished without difficulty.

FOREIGN BODIES IN THE RECTUM is a rare accident. Falling on the rung of a chair, or on fence-spokes, may result in a portion of these materials entering the rectum. The principal danger is from laceration of the bowel, uterus, or bladder. Death usually follows rupture of the latter organ.

The treatment consists in keeping the bowels quiet, relieving pain by opiates and warm fomentations to the abdomen and anus. If the mucous membrane is torn to any extent, and the injury will admit of it, the parts may be drawn together with sutures.

# CHAPTER VIII.

## *BURNS AND SCALDS.*

Varieties of Deformities produced by Burns.—Spontaneous Combustion.—Classification of Burns.—Constitutional Symptoms.—Duodenal Ulcers.—Causes of Death, etc.—Effects of Cold.—Frost-Bite.

THERE are few accidents which combine so many unnatural elements as burns and scalds. In none do we witness so much agony or such poor results from treatment.

Burns are to be dreaded in their remote results, as well as in their immediate consequences. Recovery in many cases is accompanied by hideous deformity. Severe facial burns not unfrequently leave the face twisted and distorted to such a degree as to almost destroy its semblance to humanity. The cheeks may be stretched to one side, the angles of the mouth widely separated, or the lower jaw drawn toward the shoulder, by a cicatrice of the neck. Burns of the neck may bend the head sideways, or draw it down on the chest. Where the arms or hands are burned, the cicatrices bend the joints out of place, and impair their movements. Thus the fingers may be doubled up and clinched, or the forearm flexed or strongly pronated. Sometimes the eyelids are fastened to the cheek, or drawn upward on the forehead. In the latter case the eyeballs cannot be covered or protected from irritating particles of dust; great distress results in this condition, from want of sleep. A case of this kind came under my care at Bellevue, in a female patient who suffered from a severe burn of the forehead and arm. The upper eyelid was drawn up on the forehead, and fastened above the superciliary ridge. The suffering for want of sleep was considerable. Even opiates failed to bring relief. Ordinary covering for the eye only produced irritation. Finally, as there was no integument near from which to manufacture a new lid, I dissected the old one from its attachment on the forehead, and drew it down. It was retained in its position, until the healing process became complete, by means of a fine silver wire passed through, near the free margin of the lid, carried down across the end of the nose, and fastened at the back part of the head to the other end of the wire from the opposite side. This unusual operation answered the purpose admirably. Being retained in its position for several weeks, the cicatrice was prevented from contracting so as to uncover the eye, and leave it without protection. Sleep was procured for the patient; most of the hideous deformity removed, and the old lid performed its duty once more.

Many cases of burning arise from carelessness in the use of kerosene and other explosive oils in tenement-houses. This class of burns has attained a magnitude, in point of numbers, which is truly alarming. The columns of our morning journals are seldom without the history of a victim. These accidents usually arise from filling lamps near a light, or from pouring kerosene on kindling-wood to make a brighter flame. Sometimes they are occasioned by carelessness in shutting off gas. The material escapes until the apartment is filled, and upon the entrance of a person with a light an explosion takes place, and frightful burns result.

Recovery from such burns is rare, owing to the extent of surface injured.

Dangerous burns are also produced by the contact of melted metals with the body. They burrow into the flesh, and cause great destruction of tissue, and fearful scars. Melted sugar, hot mash, boiling water, etc., when applied to the body, are not characterized by the same deep eschars which attend scalds with other substances. Their effect is superficial, but, as they sometimes extend over a greater surface of the body, they are usually as fatal as burns from flame.

The appalling phenomena of *spontaneous combustion* may be mentioned in this connection. Several cases of it are recorded by reliable observers. It takes place in persons who imbibe the worst varieties of ardent spirits. There is much diversity of opinion respecting this curious accident. Some hold that the system becomes so thoroughly impregnated with alcohol as to make ignition possible through the medium of the breath ; or, that combustible gases are generated internally, which take fire and destroy independently of external influences. The majority of investigators, however, believe that the combustion commences on the outside of the body. Thus, a person completely stupefied from alcohol may fall or lie down in the vicinity of a fire, and the flame may be communicated to his clothing. His helplessness, and the body being loaded with fat and alcohol, furnish all the materials for rapid combustion, and the unfortunate creature soon becomes a blackened, fetid mass.

In ordinary burns the danger to life varies with the seat and extent of the tissue destroyed. Burns of the thoracic or abdominal walls are attended with the greatest danger, on account of the proximity of important viscera.

A superficial burn, involving a large integumental area, is apt to prove fatal. Localized deep eschars are not particularly serious, unless important nerves or vessels are destroyed.

When the air-passages, pharynx, or œsophagus, are injured from hot liquids or steam, the prognosis is always bad.

The mortality from burns is always greater in childhood than in adults. The delicate and susceptible nervous system of the child succumbs to a burn, which would, comparatively, be of little consequence to an adult. In persons of tender years these accidents usually terminate in convulsions.

Dupuytren divides burns into six classes. Other surgeons have increased the number. For our present purposes four degrees of burns will be sufficient: The *first* includes all burns which redden the cutis and produce slight vesication. The *second* includes all cases where the true skin is either partially or completely destroyed, and bullæ or eschars of a brown color result. The *third* class includes all which extend through the subcutaneous cellular tissue into the muscular substance. The *fourth* includes those in which all the tissues of a limb are more or less involved in the destructive process.

We usually find, in burns, the first two degrees combined in the part affected. Where boiling water is spilled on the surface, the tissue is not broken up as when flame is apparent; with the worst cases the true skin is merely deprived of its cutis and reddened. Our classification, therefore, does not apply to this variety.

The immediate symptoms accompanying severe burns

may be divided into three stages, each differing in a marked degree, and giving rise to different indications for treatment. The immediate symptoms accompanying the first stage of severe burns are those of collapse. The pulse is small and feeble. The extremities are cold and clammy. There are great thirst, with difficulty in swallowing (*dysphagia*), and nausea and vomiting. The patient's countenance is shrunken, and has an expression of anxiety. Chills and rigors are present. The most prominent symptom is the intense agonizing pain. The pain is probably more acute than in any other form of injury, and oftentimes only relieved by death. This stage lasts from twenty-four to forty-eight hours, and the greatest number of fatal cases occur in it.

A *post-mortem* examination of persons who die in the first stage reveals great congestion of the brain and its membranes, serous effusion into the ventricles, and on the surface of brain. There is also marked congestion of all the internal organs.

The second stage or period of reaction is recognized by an increase in the temperature of the body, and a rapid pulse. The skin feels hot to the touch, and the tongue is brown and dry; the dryness being particularly apparent in the centre. There is intense pain in the head (*cephalalgia*), and sometimes delirium. Vomiting may also be present in this stage. The dangers in the second stage arise from inflammatory affections of different viscera. Meningitis is liable to occur. Pneumonia or bronchitis stands next in order of frequency. Inflammation of the intestines, giving rise to ulceration, is not uncommon. The inflammation usually commences in the upper portion of the small intestines. The peculiar duodenal ulcer which accompanies severe burns may take place in this period, although it is more frequently seen in the third. This ulcer is situated at the upper portion of the duodenum near the pylorus. Bowman supposes it to be caused by the extra labor thrown on the intestinal glands in consequence of suppressed cutaneous secretion. It is recognized by pain in the right hypochondrium, loose and sometimes bloody evacuations from the bowels. Usually it appears on the tenth day, but it may commence as early as the fourth.

The duration of this stage varies from one to two weeks. The *post-mortem* appearances are principally those belonging to different inflammations. If meningitis have supervened, the arachnoid will be found opaque, and studded with flakes or patches of lymph. The membrane is raised by effusion of serum into the meshes of the pia mater. The brain is congested, and the ventricles contain serum. The lungs may present various stages of pneumonia, or be simply engorged. There is congestion throughout the intestinal canal, but especially in the duodenum, and there may be ulceration.

A diminution in the febrile symptoms, and the commencement of suppuration, usher in the third stage. In severe cases, the patient's condition is similar to that of the first stage. If the suppuration be excessive, death soon ensues from exhaustion. The pathological changes are much the same as in the preceding stage, with the exception that the brain and its membranes are not so often the seat of inflammatory changes, and ulcers are more frequently found.

The most common causes of death in each period are, in the first stage, collapse from injury to the nervous system

and coma due to cerebral congestion. Second stage, inflammatory disorders, as meningitis, pneumonia, peritonitis,
etc. Third stage, exhaustion from excessive suppuration,
hæmorrhage, or peritonitis from perforation of an ulcer, and
thoracic inflammation.

The constitutional treatment varies in each period. In
the first stage the intolerable pain should be relieved by
opiates, and the patient roused from his prostration and
collapse by the free use of stimulants. And it must be
borne in mind that, when excessive pain exists, the system
can bear double doses of narcotic medicines. Two or three
grains of opium may be given to adults at short intervals,
and increased if necessary. Morphia is best administered
in solution, and, of the two liquid preparations employed,
Magendie's is the best. From twenty to thirty drops may
be given by the mouth, or by hypodermic injection. If the
preparations of opium fail, hydrate of chloral in half-
drachm doses, or anæsthetic inhalations, may be tried. Do
not let the unfortunate patient suffer, but relieve him at all
hazards.

In conjunction with narcotics, brandy may be given by
mouth or rectum. Hot bottles applied to the extremities
will be found of service. As soon as heat of the skin and
increased frequency of the pulse indicate reaction, diminish
the quantity of stimulants.

In the second stage there is an entire change in the condition of the patient. Inflammation is present in some of
the viscera. The treatment will of course vary with the
organ involved. Should the pain continue, opiates must be
administered. Stimulants may be kept up and their action
carefully watched. Antiphlogistic measures are not re-

quired. Beef-tea, broths, and other light, nourishing diet, are always beneficial, and cannot be dispensed with.

In the third stage there is great exhaustion, and efforts must be made to sustain the rapidly-failing vitality of the patient. Brandy, with or without ammonia, should be administered freely in conjunction with quinine. This valuable drug may always be employed in the treatment. Five grains every three or four hours will be sufficient. Beef-tea, raw-scraped beef, eggs, oysters, and other nutritious articles, are also essential. They may be given in all cases. If the stomach be too irritable to receive the medicine, diet, or stimulants, they can be safely given by injection.

There are three important rules to be remembered in the local treatment of burns : 1. Exclude atmospheric air. 2. Only remove the dressings when they become loosened by the discharges. 3. Prevent the contraction of cicatrices.

In simple burns which do not involve the true skin, very little treatment is necessary. The part may be kept wet by cloths dipped in water or sweet-oil. When the true skin is partially or completely destroyed, a thick layer of flour may be placed over the burned surface, and covered by cotton. Lint or cotton, dipped in a mixture consisting of equal parts of linseed-oil and lime-water (*carron-oil*), can be used instead of the flour. Some envelop the burnt part in cotton saturated with sweet-oil alone, and others apply a solution of nitrate of silver first, then cover the lint with cotton. I have seen the best results from the employment of flour and *carron-oil*, and prefer them over all others. Whatever dressing is employed, it should not be disturbed until separated by the exudation underneath, or unless foul odors arise. In changing, every particle should be carefully removed, and the parts thoroughly washed with some disinfectant liquid, such as

R. Acidi carbolici . . . . . . . 3j.
Aquæ . . . . . . . fl. ℥ viij. M.

This solution may also be sprinkled on the dressings and bedclothes.

When granulations grow above the surface, the sore will not heal; applications of nitrate of silver and strapping with adhesive plaster will then be required.

During cicatrization, the great tendency to contraction and deformity must be counteracted by splints or bandages, and parts supported in their normal position until the healing process is completed. The hideous deformities which arise from the contractions of cicatrices are sometimes remedied by surgical procedures. No special rules can be laid down for those operations, as each one has its own separate requirements, and the common-sense of the surgeon must alone be the guide.

### EFFECTS OF COLD.—FROST-BITES.

Cold is a valuable therapeutical agent in many diseases. Cold shower-baths or ordinary cold-water baths have a stimulating effect on the system, invigorating both the mental and physical forces. A dry cold atmosphere is also an efficient agent in maintaining the vital powers at a normal standard, and in destroying or keeping in abeyance injurious miasm.

Exposure of the body to intense cold results in a local or general loss of vitality. It produces a feeling of depression, a disturbance of the mental faculties, and a great desire to sleep, which, if indulged in, soon increases until a state of profound coma is reached which may end in death. The desire to sleep is beyond the control of the sufferer, and it is here that the great danger lies. If the power of resistance, or an appreciation of the danger were felt, the person exposed might be enabled to resist until assistance was obtained. When the coma is developed, it is almost impossible to arouse the patient.

The comatose condition is brought about by congestion of the brain. The intense cold propels the blood from the surface to the internal organs. The functions of the brain, in common with those of other organs, are interfered with by the pressure of the accumulated blood, and insensibility supervenes. It is also probable that an accumulation of carbonic acid takes place in the blood owing to the diminished respiratory movements, and through its narcotic effect assists in producing the coma. Fatigue and intemperance are two great auxiliaries in making the system susceptible to the effects of cold. Persons who have been overworked, or who have imbibed freely of alcoholic beverages, succumb readily to cold. Temperate men resist long exposure to a low temperature.

The condition of the atmosphere modifies the effect of cold. Thus a much lower temperature can be borne when the atmosphere is still than when the wind is blowing. When a breeze exists, the warm stratum of air nearest the body is removed rapidly, and cold air takes its place; there is consequently more heat abstracted from the body than in the former condition. Air is a bad conductor of heat, and these warm strata afford a certain amount of protection, and lessen the demand for a higher temperature.

When only a portion of the body is exposed to the cold,

as the eyes, ears, nose, etc., there is a local loss of vitality. The part becomes pale and bloodless, and is devoid of sensation. If the vitality is only partially destroyed, a condition arises which is known as frost-bite; where the exposure has been long continued, and the life of the part totally destroyed, gangrene rapidly ensues. Little or no pain is experienced until recovery begins, and the circulation is renewed. The pain is intense, and always the forerunner of more or less inflammation. The parts become red, swollen, and hot, and the cuticle peels off. Resolution may occur in a day or two, or the inflammation may continue until sloughing or gangrene takes place.

Extreme degrees of cold and heat have analogous effects. In both the vitality is destroyed, and in both there are subsequent inflammation and sloughing of tissue, with constitutional disturbance.

*Treatment.*—A person suffering from frost-bite should be placed in a cold room. The part frozen may then be rubbed with snow, or ice-water poured on it, until sensation begins to return. The occurrence of stinging pain, with a change in color, is a signal to stop all rubbing or other measure which might excite inflammation. Cloths wet with ice-water may then be applied to the part. If the inflammation extend to the deeper tissues and suppuration occur, the cloths can be wet in a solution of carbolic acid and ice-water, and the application continued. When gangrene sets in, amputation is generally necessary.

In cases where the constitutional effects of cold call for treatment, general stimulation is necessary. Brandy and ammonia are to be given internally, while the body should be briskly rubbed with the hands and warm flannel.

# CHAPTER IX.

## STRANGULATED HERNIA.

Causes and Symptoms of Strangulation.—Ileus.—Volvulus.—Taxis.—Operations
for Inguinal and Femoral Herniæ.

THE escape of any viscus from its natural cavity is called
a *hernia*. The term is in a measure restricted to the protrusion of a portion of intestine or omentum from the abdominal cavity. The affection is of common occurrence.
In ordinary cases it is attended with little inconvenience
or danger. If, however, a constriction takes place at the
neck of the hernial sac, which cuts off the circulation of
blood, and obstructs the passage of fecal matter through
the intestines, the patient's life is at once in jeopardy. The
portion of intestines so constricted is termed a strangulated
hernia.

Hernial protrusions usually occur at the inguinal or
crural canals ; but they may pass through the umbilicus,
or other part of the abdominal walls.

A hernia may become strangulated : 1. From the addi-
tional protrusion of intestines or omentum into the sac
during the act of straining, or other violent exertions which
bring the abdominal muscles into violent action.

2. Thickening of the sac or its contents by cell-pro-
liferation, or deposit of adipose tissue.

3. Contraction of bands of fibrine over the neck of the sac.

4. Spasmodic contraction of the muscular fibres at the same point.

5. Contraction of the ring, from growth of new tissue.

All of these causes may combine to induce strangulation. In inguinal hernia the constriction is usually situated at the internal or external abdominal rings. In femoral hernia it may be at the crural ring, or the saphenous opening.

The strangulation is first manifested by pain over the hernial tumor. The pain increases in intensity, and rapidly spreads to other portions of the abdomen. Soon there are nausea and vomiting. The vomited materials consist first of the contents of the stomach, and then of stercoraceous matter. The bowels are obstinately constipated. Cathartics fail to influence them. The pulse is rapid, increasing in feebleness as the strangulation continues. The abdomen is tympanitic, and pressure at any part is attended with great pain. This indicates the extension of the peritoneal inflammation. Finally, the extremities become cold and clammy, and the pulse can scarcely be distinguished at the wrist. All the signs of collapse are present, and death rapidly ensues unless the strangulation be relieved. When collapse sets in, operative measures are of little avail.

In all cases where a patient is vomiting, and complaining of pain in the abdomen, an examination should be instituted for hernia. Fatal mistakes are made by neglecting this precaution, and the sick person treated for colic and indigestion. At the same time it is well to avoid the other extreme, and take care not to cut into an inflamed bubo, or an inflamed incarcerated hernia, on the supposition that

8

strangulation exists. These things are occasionally done even by men of standing. If obstinate constipation and vomiting of fecal matter exist, there is little room for mistake; neither of these will be connected with a bubo or inflamed hernial sac.

The intestines may be constricted without leaving the abdominal cavity. Portions of the colon may twist upon themselves (*ileus*), in such a manner as to cut off the circulation. The twisting is usually found at the sigmoid flexure. It is recognized as a prominent tympanitic tumor over the part affected, and by accompanying signs of strangulation.

A portion of intestine may become invaginated or inverted (*volvulus*), like the finger of a glove doubled in, and occasion all the symptoms and danger of strangulation. Volvulus may occur at any age, but it is most common in childhood. It occurs suddenly, with pain located at the point of constriction. In addition to the ordinary signs of strangulation, there are frequent desire to go to stool, and discharges of blood and mucus from the bowels. The invaginated part may slough off—the two ends of the intestines unite, and the patient recover. If allowed to remain until sloughing occurs, a favorable termination is not likely to ensue.

*Treatment.*—The injection of air or fluids into the intestines is highly recommended in volvulus. The injected material, by distending the gut, forces up the invaginated part. Some recommend cutting down upon the intestines at the part where the pain exists, and drawing out the inverted intestine. A similar course may be adopted in the treatment of ileus.

In ordinary strangulated hernia, efforts should be made

to reduce it by manipulation (*taxis*). The muscles are first relaxed by opium, hot baths, or anæsthetics. The thigh is then partially flexed and adducted, and the body of the patient raised in bed. Firm pressure is then made with the right hand on the tumor, while the left is placed at the neck of the sac, to keep it from bending or doubling upon itself in the reduction. Taxis must not be kept up too long, or performed with violence. Great pressure may force the hernia, constriction and all, back into the peritoneal cavity. Such an accident complicates matters. Should the manipulations be without avail, the constriction must be removed at once by an operation. The patient is first put under the influence of an anæsthetic. If the hernia be of the oblique, inguinal variety, an incision is made through the integument in the long diameter of the sac. The succeeding layers are opened on a director. They are in order from without, inward—two layers of superficial fascia, intercolumnar fascia, cremaster muscle, infundibuliform fascia, subserous areolar tissue, and peritonæum. When the tissues are thickened, a greater number of layers may be made by splitting up the fascia with the director. These layers are not always recognizable. Some surgeons repudiate them altogether, and rely upon the appearance of the sac or its contents as a guide. The peritonæum is recognized (provided it is not thickened by inflammation) by its tension, and the arborescent arrangement of its blood-vessels. If the peritonæum cannot be recognized before, it may be after it is cut through, by the escape of dark-colored serum, which generally exists inside the sac. The intestines are known by their dark color and polished surface. When the intestine is exposed, the little-finger of the left hand is passed up to the part of stricture which can be felt like a "hard, bony ring" at the neck of the sac. A hernia-knife, or an ordinary bistoury, with its point protected by adhesive plaster, is then introduced on its flat surface, be-tween the nail of the little finger and the constriction. When it has passed under, the edge of the blade is turned up, and the stricture cut directly upward. By cutting in this direction, the epigastric artery, which runs up between the two rings, is avoided.

If the intestine is in a fit condition to return to the ab-domen, it will change color soon after the stricture is re-lieved. In this case it is returned slowly—the part which came out last being replaced first. Should gangrene have set in, there will be a fetid odor, the intestine will be of a dark-gray color, and may crepitate on pressure, from the presence of putrefactive gases in the walls. The gangre-nous portion is to be removed, and an artificial anus made by sewing the cut ends to the edge of the opening.

In direct inguinal hernia, the layers are somewhat dif-ferent, but the operation is precisely similar. Instead of the cremaster muscle, the conjoined tendon of the internal, oblique, and transversalis muscles is substituted, and the in-fundibuliform fascia is replaced by the fascia transversalis.

In operating for femoral hernia, a crucial or a T-shaped incision is made—the first one in the long diameter of the sac, parallel with Poupart's ligament, and the second meet-ing the first at right angles. The layers to be divided are: The integument, superficial fascia, cribriform fascia, crural sheath, septum crurale, subserous areolar tissue, and perito-næum. The stricture is divided by cutting upward and in-ward. In order to avoid cutting the obturator artery, which occasionally runs along the inner edge of Gimbernat's ligament—the edge of the knife may be blunted prior to the operation. When this is done, the artery will be pushed before the knife, instead of being wounded.

Taxis is employed in femoral hernia, by first flexing the thigh, rotating it inward, and pressing the protrusion downward, backward, and then upward.

# CHAPTER X.

## COMA.

Coma from Compression of the Brain — Embolism — Uræmia — Alcohol.
Hysteria—Epilepsy.—Concussion.

A SUSPENSION of cerebral activity and unconsciousness is the common sequence of many abnormal changes. It may result from structural lesions in the brain, or from the effects of poisonous substances carried to that organ by the blood. It may arise from a deficient supply of healthy blood to the nerve-tissue, as in syncope, or from defective aëration of the blood, as in asphyxia.

Coma which arises from cerebral lesions, or from the circulation of urea or alcohol in the blood, will be considered in this chapter.

By the term coma we mean a state of partial or complete insensibility—a suspension of the ordinary powers of sensation and volition, accompanied by stertorous breathing. As this condition is merely a representative of diverse disorders, a just appreciation of the cause of each variety is essential to effective treatment.

The causes of coma are: 1. Pressure on the brain-substance, from extravasated blood, depressed fracture of·

cranial bones, and inflammatory products; 2. Anæmia of the brain, as in embolism and thrombosis; 3. Blood-poisons, as urea, alcohol, etc.; 4. Epilepsy; 5. Hysteria.

Extravasation of blood on the surface of the brain is usually the result of external violence. When it occurs in the substance of that organ, it proceeds from a diseased condition of the cerebral blood-vessels. They may be affected by simple fatty or atheromatous degeneration. According to Virchow, the latter commences as a low grade of inflammation in the lining membrane of the artery There is a slight exudation between the inner and middle coats, and subsequent softening and breaking down of the different layers. In the *débris* of disintegrated tissue we find fat, cholesterine, calcareous salts, and albumen. If there is any increased action of the heart while this morbid change is in progress, the weakened walls of the capillaries are liable to give way, and allow the blood to escape.

The extravasation is most frequently located in the corpus striatum and optic thalamus, portions of the cerebrum possessing the greatest vascularity, and therefore more liable to the affection. When the blood is found on the surface, the meningeal arteries are generally ruptured, the middle meningeal more frequently than the rest.

The coma which arises from laceration of diseased arteries, in most instances, is sudden in its development. In very rare cases there are premonitory symptoms, appearing in the shape of slight facial paralysis, twitchings of the muscles, local points of anesthesia in the extremities, and bleeding from the nose. In some cases the delicate vessels of the retina rupture, and produce blindness. This occurred in the case of the late Dr. George T. Elliot. He suffered from retinal apoplexy several months previous to the extravasations in the brain which ended his life.

When the attack is sudden, the patient falls to the ground insensible. The face presents a congested appearance; one pupil may be dilated and the other contracted, or both may be dilated. They will not act readily to the stimulus of light. If the clot of blood involve both sides of the pons Varolii, both pupils will be contracted. Strabismus exists in many cases. The respiration is labored and stertorous; with each expiration the cheeks are puffed out, as in the act of blowing. The peculiar noise, or stertor, accompanying the respiratory movement, is due to a partial paralysis of the soft palate and pillars of the fauces. The pulse is slow and full; the integument is warm and moist, but there is no increase of the natural temperature of the body. Paralysis of one side (*hemiplegia*) is usually present. When both sides are paralyzed, the extravasation will be found in the *pons*. In the face the paralysis is indicated by a drawing down of the angle of the mouth on one side, and a diminished movement on the other, or perhaps with inability to close the eye (*lagophtnalmus*).

If the clot involve the crura cerebri so as to press on the third pair of nerves, there will be inability to open the eye (*ptosis*), divergent strabismus, and dilatation of the pupil on the side opposite to the general paralysis. Paralysis of the face is, in the majority of cases, on the same side as the hemiplegia. Paralysis of the extremities is seen in the different effects produced by counter-irritation, one limb moving more than another when pounded or pricked. An instrument called an æsthesiometer is now employed to ascertain the different degrees of sensibility existing in various parts.

The sphincter muscles which guard the rectum are also paralyzed, and the fæces are passed involuntarily. The orifice of the bladder is guarded by elastic fibres, which retain the urine when the sphincter of that organ is paralyzed. The coma which follows external violence presents similar symptoms, whether connected with depressed bone or extravasated blood.

There are exceptional cases of cerebral extravasation which do not exhibit these dangerous characters for two or three days succeeding the injury. The patient may have been treated for a slight scalp-wound, without any suspicion of the real nature of the lesion. He may pursue his usual avocations with little trouble until he suddenly sinks into a state of coma, with the signs of compression plainly manifested.

A *post-mortem* examination in these cases shows that the effused blood is located principally at the base of the brain, and that it is connected with fracture of the base of the skull.

When coma supervenes three or four days after an injury, accompanied by an increase in the pulse and temperature, the pressure of inflammatory products, such as serum, lymph, or pus, may be suspected. The formation of pus, or the occurrence of pyæmia, is announced by severe rigors.

An injury to the head may be followed by entirely different symptoms from those previously described. The patient may have concussion of the brain without compression. There is loss of consciousness in both; but, in concussion, the patient is more easily roused, the face is pale, and the surface of the body cold. In compression, the face is flushed and the body warm. Stertorous breathing characterizes the latter; in the former the respiration is natural or sighing. The pulse, in concussion, is small and rapid; in compression, it is slow and full. The pupils are generally contracted in concussion, while in compression they are dilated. The condition of the pupils, however, should not be relied on in diagnosis, as it is subject to much variation. In compression of the brain, there is usually paralysis, which alone is sufficient to distinguish it. In rare instances, compression and concussion are combined. In such cases, remedial efforts are mainly directed to relieve the former.

It is necessary to diagnose apoplectic from uræmic coma. With the latter there is usually a history of Bright's disease of the kidneys, œdema of the lower extremities, a pale, waxy countenance, and albumen and casts in the urine. In the former, these signs are usually absent. Apoplectic coma is attended by paralysis of one side of the body, and the pupils are irregular. In uræmia there is no paralysis, and both pupils are dilated. The temperature of the body is said to be higher in uræmia than in apoplexy, but this cannot be depended on in diagnosis. When the urinous odor of the perspiration exists, we have further evidence of uræmia.

*Treatment.*—Very little can be done to relieve the coma which results from the rupture of diseased arteries. If the patient is plethoric, the abstraction of a few ounces of blood from the arm may prevent further extravasation. Authorities differ as to the utility of this measure. The after-treatment consists in the prevention of inflammation and the administration of medicines, which assist in the absorption of the clot. For the latter purpose, iodide of potassium may be administered in doses of from five to ten grains three or four times each day. If the stomach is disordered, or an eruption of the skin is produced by its use, it should be discontinued. If inflammation be apprehended, mustard-poultices may be applied to the nape of the neck and to the feet, and the bowels should be thoroughly moved by an active cathartic. Croton-oil and elaterium are the most efficient.

If the extravasation proceed from a blow or fall on the head, the operation of trephining can be performed in one of two places, viz. : near the course of the middle meningeal artery on the side opposite to the paralysis, or directly underneath the point where the injury was inflicted. A crucial incision is made through the scalp, which is turned back and the bone exposed. The skull is then cut carefully through with the trephine. If the blood is found between the dura mater and the bone, it is readily removed. If the membrane swells up through the opening, and there appears to be blood underneath, an incision can be made through it to allow its escape. After the operation, the wound is covered, and simple water-dressings applied. The usual remedies, previously mentioned, to prevent inflammation, are then employed.

When the coma arises from depressed fracture of the skull, trephining is resorted to, or the depressed bone is raised by an elevator.

COMA FROM EMBOLISM AND THROMBOSIS.—Inflammation of the valves of the heart and atheromatous degeneration of the aorta are attended with the formation of fibrinous masses, which project beyond the natural dimensions of the artery and valve, and are liable to be washed away by the current of blood. These small particles may be carried to the brain and plug up one of the cerebral arteries, cutting off the supply of blood from that portion. The artery most frequently involved is the left middle cerebral. The plug is called an embolus.

Diminished action of the heart, with loss of elasticity in the walls of the vessels, may predispose to the formation of a clot of blood (*thrombus*) in them. The supply of circulating fluid is cut off as in the former case, and anæmia of the part results. Either of these accidents, taking place in the brain, may produce *coma*. In some cases this is gradual, in others the attack is sudden. The coma differs very little from that which depends upon cerebral extravasation. In coma from plugging of arteries, the face is usually paler than in cerebral extravasation, and there is with it some disease of the mitral or aortic valves. Another important point in the diagnosis is, that consciousness is restored more rapidly in the former (often within two or three days), and that the paralysis is not so persistent.

*Treatment.*—In these cases we can only wait for developments. If softening of the brain be apprehended, stimulants and tonics are indicated. Some recommend the administration of ammonia to absorb the clot of fibrine. Its remedial action is, however, questionable.

URÆMIC COMA results from the same poison which induces uræmic convulsions. Frerichs developed the fact that these phenomena were caused by the accumulation of urea in the blood, and its subsequent change into carbonate of ammonia. Spiegelberg, a later investigator, has fully confirmed these views by a series of carefully-conducted experiments.

Urea is produced by the decay of nitrogenized tissue. It is eliminated by the kidneys. When these organs are diseased, its channels of escape are almost wholly closed, and it accumulates in the blood. There it is decomposed, one atom of urea taking two atoms of water from the blood, and forming by this combination carbonate of ammonia.

Uræmic coma occurs during the progress of Bright's disease of the kidneys, and may have all the symptoms of that affection connected with it. The patient's face has a pale, waxy look. There is dropsical effusion in the cellular tissue of the lower eyelids, and behind the ankles, or over the whole of both lower extremities. The urine is of low specific gravity. It contains albumen and casts. Preceding the coma there are headache, dimness of vision, and vomiting. The patient passes into a somnolent condition, which hourly increases, until a state of profound coma is reached. Sometimes the coma is preceded by a convulsion, without other premonitions. This is observed especially in the small contracted kidney.

The coma is accompanied by a certain amount of stertor. The pupils are dilated, but not irregular. The pulse is more rapid than usual, and lacks firmness. The temperature is sometimes slightly increased.

Poisoning by belladonna presents some similarity in its symptoms to uræmic coma. The pupils in both are widely dilated, and the insensibility is profound. The history of the case, and the absence or presence of signs of Bright's disease, determine the diagnosis.

*Treatment.*—Our principal efforts in all cases is to eliminate the poison from the system, through the medium of the skin and bowels, with diaphoretics and active cathartics.

Mix equal parts of croton-oil and ordinary sweet-oil, and apply four or five drops of the mixture to the back of the tongue. This can be done by moistening the end of a pencil or pen-handle with the oil, and wiping it on the back of that organ. It is not well to use the croton-oil undiluted, on account of its irritating properties. The dose should be repeated in three-quarters of an hour, if free evacuations from the bowels do not follow. If preferred, elaterium may be administered in quarter-grain doses every hour until a like effect is produced. In connection with the internal medication, profuse sweating should be produced by means of hot-air baths. Bottles of hot water and warm blankets, applied to the surface, answer the same purpose. The sweating may be kept up for a considerable time without injury, but the action of cathartics must be guarded, especially if the constitution be much weakened. In ordinary cases, this treatment should be persevered in until consciousness is restored. Prof. A. L. Loomis has lately employed morphia in uræmic coma. He administers it hypodermically, and with good success. Subsequently the action of the skin may be kept up by warm baths and mild diaphoretics. Tonics and nourishing diet are also necessary. To sustain the action of the kidneys, and at the same time to support the strength, the following may be given in teaspoonful doses four or five times a day:

B. Hydrg. bichloridi . . . . . . . gr. j.
 Tinct. cinchonæ comp. . . . . fl. ℥ iv. M.

The internal administration of benzoic acid was at one time proposed as an antidote for the poison of urea; experiments, however, did not warrant a continuance of its

use. When uræmic coma is the result of acute inflammation of the kidneys, the treatment varies. In addition to
the ordinary remedies, the application of wet or dry cups
over these organs is required, and is generally followed by
great results.

RUM COMA.—When large quantities of alcohol are taken
into the system, a state of insensibility is induced which in
certain particulars resembles the other varieties of coma.
The comatose or "dead drunk" patient lies insensible,
breathing heavily. The respiration has more of the character of a snore than of a true stertor. The pupils are
regular and act to light. Sometimes they are dilated.
In the early part of the coma the pulse is soft and increased in frequency, but afterward becomes slower. The ·
breath usually smells strongly of alcohol. Too much reliance, however, must not be placed on this sign until the
history of the case is examined into, for, in cases of sudden
insensibility, by-standers are in the habit of administering
stimulants. The patient usually has been drinking freely
for some time, and the stupor appears gradually, preceded
by a staggering gait, and other signs of drunkenness. Coma
due to compression of the brain may be excluded, if there
is no paralysis or irregularity of the pupils, or complete
coma. From uræmic coma it is diagnosed by the absence
of œdema of the eyelids and lower extremities, of albumen
or casts in the urine, or urinous odor in the perspiration
Besides, uræmic coma is profound, while coma from rum
is only partial. If the patient had a convulsion previous
to the coma, and no signs of Bright's disease present,
the case might readily be mistaken for true epilepsy.
Our main reliance under such circumstances must be the

history of the case and the surroundings of the patient. If the tongue has not been bitten, and there is a history of a spree, we may then exclude epilepsy.

*Treatment.*—If an emetic of mustard can be administered, and the stomach emptied, much good will result. Subsequent applications of cold water to the head and chest will be beneficial.

HYSTERICAL COMA is one of the manifestations of the hydra-headed nervous affection hysteria, a disease peculiar to nervous women. Scientific investigation has not yet reached the morbid changes which occasion the disease. Its real nature is still in the dark. We know that it is characterized by a morbid sensitiveness, a tendency to imitate disease, and that it is to a certain extent under the control of the will, but farther we cannot go.

The patient imagines she has a disease, but the practised eye detects the counterfeit. She may simulate paralysis, and remain in bed for months. All the pains, aches, and diseases, which "flesh is heir to," may be represented and imitated without limit, and yet these unfortunates cannot be charged with fraud. The case of a young hysterical patient, who was at one time in Ward 24, Bellevue Hospital, furnished an excellent example of this class. On her admission, she was placed near a patient in the last stages of Bright's disease. In a few hours afterward, I found her suffering from nearly every prominent symptom exhibited by her dying neighbor. The condition lasted for a few days, when the ambitious young woman developed the signs of peritonitis, and managed to keep them up for two or three weeks. Subsequently, she passed to the care of another house-physician, and I lost sight of her. In another ward of the same hospital was a young Irish girl who suffered from retention of urine. The catheter was regularly employed for several days before the real nature of the disease was discovered. Her will, or her disease, enabled her to remain three days without passing water. At the end of that period she relieved herself naturally, and continued to do so afterward. The same patient afterward developed paralysis of the lower extremities, which lasted several months. Temporary recovery took place during a thunderstorm. The noise alarmed her so that she forgot her paralysis and sprung out of bed. It returned again in a milder form, but gradually wore away. When discharged from the hospital, she was entirely cured.

Hysterical coma is a comparatively rare manifestation of the disease. It is often preceded by general excitability, and by spells of violent laughter and crying without assignable cause. There is often a sense of choking (*globus hystericus*), due to contractions of the œsophagus, from below upward. It gives a feeling as if a ball were rising in the throat. Previous to the coma there may have been a convulsion, but it is not always an accompaniment.

The patient, during the attack, lies motionless, and to all appearance unconscious. The breathing is natural. There is no lividity or other unnatural condition of the face. An examination of the eyes will show that the patient sees all that is passing around her, and that the pupils act to light. The pulse is natural in all respects. The absence of stertorous breathing, insensibility, and irregularity in the pupils, suffices to show that there is no compression of the brain or other serious affections.

*Treatment.*—For hysterical coma, the cold douche is the

9

best known remedy. Two or three pitchers of cold water, poured from a height upon the face, will generally suffice to bring about a recovery. The after-treatment consists in developing self-control, sustaining the general health with fresh air and good food, the removal of any existing disease of the generative apparatus, and the administration of anti-spasmodics, as musk, valerian, assafœtida, etc.

EPILEPTIC COMA follows an epileptic convulsion. The insensibility is never complete. Blood may collect on the lips. There is laceration of the tongue. The sudden oc-currence of the convulsion when the patient is in good health otherwise, and the complete recovery when the attack has passed away, serve to distinguish this disease in all cases. (*See* article on Epileptic Convulsions.)

*Treatment.*—Epileptic coma does not require treatment. To prevent a recurrence of the convulsion, bromide of po-tassium can be given. Ten grains, four times a day, will be enough for an adult.

### CONCUSSION OF THE BRAIN.

Concussion of the brain may be defined as a shaking to-gether of the contents of the cranial cavity, with more or less contusion of the brain-substance, and attended by par-tial or complete unconsciousness. The injury may be pro-duced by direct blows upon the head, or by jumping from a height and alighting on the heels, the force in the latter case being transmitted through the spinal column.

In some cases the most careful examination of the brain after death fails to detect signs of contusion. In the major-ity, however, minute points of extravasation, discoloration, and softening of small portions of the nerve-substance, are found.

Millar, Wood, and others, divide concussion into three stages: 1. That of insensibility; 2. Reaction; and 3. Excessive reaction or inflammation. The symptoms attending the first stage vary with the amount of concussion. In typical cases, the patient falls unconscious after receiving the injury. The skin is pale and cold, and the pulse small and rapid. Respiration is natural or sighing. The pupils are contracted, or one may be contracted and the other dilated. The sphincter muscles are not often interfered with.

In the second stage, the patient vomits and shows evidences of returning consciousness. The pulse becomes stronger, warmth returns to the body, and slight color to the lips and cheeks. If this reaction be excessive, showing a tendency to inflammation, the third stage is ushered in. The skin becomes dry and hot, and there is considerable headache. The pulse rises, and is firmer than during the preceding stages. Finally, if the case progresses unfavorably, all the signs of meningitis are manifested, such as intolerance of light, intense headache, contracted pupils, subsultus tendinum, delirium, and finally coma. The differential diagnosis between compression of the brain and concussion has already been given.

In many instances, the concussion is extremely slight, lasting but a few moments. This is the case where the patient is merely stunned, and the effect soon passes away. In other cases, the concussion is so great as to cause instant death.

*Treatment.*—If there be collapse, hot bottles and blankets are to be applied to the extremities, and the circulation stimulated by friction with the hands. Diluted enemata of brandy and ammonia are also serviceable. All stimulating efforts must cease as soon as reaction returns. Should inflammation set in, the ordinary antiphlogistic treatment, previously referred to, will be necessary.

# CHAPTER XI.

## SYNCOPE.

Syncope from Hæmorrhage.—Thrombi in the Pulmonary Vein.—Anæmia.—Mental Emotion.—Blows on the Epigastrium.—Collapse.

THE normal performance of every function depends on an adequate supply of healthy blood. The delicate machinery ceases when the proportion to each part is not commensurate with its demands.

The continuous pulsatory movements of the heart propel the blood into the vessels which carry it to all parts of the body. A partial or complete cessation of the action produces a condition known as syncope, or fainting. This is characterized by unconsciousness, and by suspension of the powers of volition.

The regular contractions of the heart depend upon several conditions: 1. A sufficient and regular supply of blood, which exercises a stimulating effect on its fibres; 2. A normal proportion of the necessary ingredients in the circulating fluid; 3. A healthy state of the brain and of the nerves and sympathetic ganglia which supply the heart; 4. A special irritability possessed by the muscular fibres, which causes its contractions to continue even when all connection

with the body has been severed, and the extraneous sources
of stimulation removed.

This innate power is, for want of a better name, denomi-
nated irritability. Of its nature we are totally ignorant.
In cold-blooded animals it is particularly noticeable. Any
morbid change, which directly or indirectly disturbs the conditions spoken of, is liable to induce syncope.

Syncope is produced by excessive hæmorrhage. This,
however, when not too prolonged, is rather of benefit than
otherwise. The cessation in the movements of the heart
allows the blood to coagulate in the bleeding vessels, and
prevents the possibility of hæmorrhage when the circulation is renewed.

Thrombi in the pulmonary vein causes fatal syncope by
preventing the blood from passing through the lungs to the
left side of the heart, and by producing distention of the
right auricle and ventricle.

Syncope arising from a deficiency in the ordinary stimulating ingredients of the blood is witnessed sometimes in
anæmia, and in chlorosis. In these diseases the watery
portions of the blood are increased, the red corpuscles are
diminished, the circulation being at all times exceedingly
feeble. In leucocythæmia, where there is a very great excess
of white corpuscles, and in phthisis, where there is much
general deterioration of the blood, sudden failure of the
heart's action is likely to occur after rapid exertion.

Syncope likewise results from mental emotions, such as
sudden joy, anger, grief, etc. These act in some peculiar
and unknown manner upon the nerves of the heart, sus-
pending their influence. In some cases the emotion has
been so great as to destroy life.

Anæmia of the brain and concussion are attended with syncope. Blows on the epigastrium may injure the solar plexus, and cause a fatal reflex paralysis of the heart. The cases of sudden death from drinking cold water while perspiring are similarly accounted for.

Sedatives may induce syncope if the doses are large or too frequently repeated. The majority of sedatives, such as tobacco, colchicum, antimony, prussic acid, etc., act by diminishing the nerve-force. Some consider that digitalis acts on the heart as a tonic, and not as a sedative. It is hard to harmonize with this theory the authenticated cases of syncope, or collapse, following its use in the usual medicinal doses.

Chloroform, when administered to debilitated individuals, may act directly upon the nerves of the heart, and cause paralysis of that organ. Chloroform usually kills by acting through the lungs and producing asphyxia, or through the brain, causing coma.

Severe burns, crushed limbs, surgical operations, etc., are sometimes followed by sudden partial suspension of the functions of the nervous system, and diminished action of the heart, which is commonly known as shock or collapse. Although in many essential points resembling ordinary syncope, there are important differences which distinguish them. The duration of syncope is more brief. The patient either dies suddenly or recovers rapidly. Collapse is prolonged. Syncope is attended with unconsciousness and loss of voluntary motion. In collapse the patient is not completely insensible, the mind is to a certain extent clear, and the power of voluntary movement remains.

Other varieties of syncope arise from disease of the heart or its coverings. Among them are fatty degeneration of the muscular fibres, angina pectoris, and pericarditis, with effusion.

Persons of delicate frame and sensitive nervous organizations are most subject to syncope. Women are affected more frequently than men. Feeble women, with uterine disorders, will faint from slight injury, or any unusual mental excitement.

The symptoms of syncope are clearly marked. The patient is conscious of a sinking sensation in the epigastric region, and about the heart. There are dizziness, dimness of vision, and ringing in the ears (*tinnitus aurium*). The features are pinched, and the lips and cheeks are pale and cold. The pulse, at first small and fluttering, is at last imperceptible. An impulse can scarcely be recognized in the præcordial region. There is also partial or complete unconsciousness. Respiratory movements may cease altogether, or a spasmodic, irregular sighing is present.

The attack lasts from a few seconds to two or three minutes. It is very rarely prolonged beyond two minutes. Resuscitation would not be possible if the heart's pulsations were absent for five minutes (*Walsh*).

Recovery is announced by attempts at swallowing, by sighing, movements of the body, restoration of warmth and color to the cheeks, and a return of the radial pulse. In some cases the attack may terminate with nausea and vomiting.

Although in most cases syncope is easy of recognition, mistakes are sometimes made and erroneous opinions given. It is therefore well to consider the morbid states for which it may be mistaken.

There is a class of persons called *malingerers*, who, from

sordid or other motives, feign various forms of illness, and syncope is sometimes simulated. Prostitutes or disorderly characters, in order to escape detention in the station-house, or a subsequent visit to Blackwell's Island, work on the sympathies of the police official, until a carriage is ordered, and they are conducted to the hospital. Once there, unless the doctor in attendance is particularly disgusted with the performance, the patient will likely be discharged the next day without trouble. These cases are readily recognized by the fact that the pulse is beating with its accustomed fulness and regularity, that the temperature of the body is normal, and that an announcement of an intention to draw blood from the arm, or shave the head and apply ice, is followed by an avowal of the patient that she is much better, and will not require further treatment.

Ordinary syncope is readily distinguished from hysterical stupor by the fact that the patient has not lost consciousness, nor is the action of the heart or pulse specially altered.

Poisoning from carbonic acid gives a dark, livid color to the countenance, the insensibility is continuous, and the pulse can be felt in the wrist. Poisoning from urea, or Bright's disease, is diagnosed by the accompanying dropsical swelling of the lower limbs, urinous odor, and the presence of casts and albumen in the urine.

A person in a state of deep syncope may be considered dead, but if the characteristic signs of death are understood, little difficulty will be experienced in making a correct diagnosis. (*See* article on Asphyxia, page 147.)

*Treatment.*—In mild cases, where the patient is only partially unconscious, stimulating inhalations of eau-de-cologne, vapor of ammonia, sprinkling the head and face with cold water, or placing the patient in a cold draught of air, will suffice to restore sensibility.

Where there is complete unconsciousness, more urgent measures will be necessary. In all cases, the patient should be placed in the recumbent position, with the head lower than the shoulders. This is done in order that the blood flowing toward the cerebrum may have the assistance of gravitation, and also to accelerate the current travelling from the lower extremities toward the heart. All superfluous clothing should be removed from the chest and throat. Collars, neck-ties, and other articles which constrict the neck, hinder recovery. The stimulating inhalations of ammonia, etc., are of little avail in complete syncope, for there is scarcely any respiratory movement; the nostrils, however, may be moistened with the liquid. Cold water, thrown violently in the face, or sprinkled forcibly on the chest, striking the palms of the hands, and rubbing them rapidly, are efficacious in all cases. An efficient remedy is to dip a plate in hot water and place it over the epigastric or præcordial regions; either place will answer. All these methods may be combined in the treatment of individual cases. Should they fail, galvanism may be carefully tried. Too much is worse than too little. One pole of the battery may be placed at the upper part of the spinal column, and the other moved up and down over the sternum and præcordia. The poles may also be applied along the course of the spinal accessory nerve. The action of the heart has in some cases been renewed by exciting the spinal accessory and the four upper cervical nerves (*Valentin*).

The treatment of syncope resulting from excessive hæmorrhage has been discussed in a preceding chapter.

# CHAPTER XII.

Respiratory Apparatus.—Effects of Non-aëration of Blood.—Strangulation.—Compression of Thorax.—Inhalation of Poisonous Gases.—Signs of Death.—Drowning.

THE pathological changes arising from defective aëration will be better understood if we glance briefly at the processes which regulate the supply of oxygen, and the elimination of carbonic acid. To describe in detail these important phenomena would lead us beyond the prescribed limit of this work. We must confine our attention to such as have a special bearing upon the morbid actions in question.

The respiratory apparatus comprises the larynx, trachea, bronchi, and lungs. The lungs, the heart, and great vessels, are contained within the cavity of the thorax or chest. A large, flat muscle, called the diaphragm, forms the floor of this cavity and separates it from the abdomen. Each lung is composed of bronchial tubes, air-cells, vessels, and nerves. The bronchial tubes commence at the termination of the trachea. They divide and subdivide, becoming smaller as they pass in, until they terminate with a diameter of $\frac{1}{50}$ of an inch in the intercellular passages or bronchioles. Around these passages and terminal bronchi, the air-cells are clustered in a manner similar to the arrangement of " leaves on a tree-branch." These cells measure from $\frac{1}{70}$ to $\frac{1}{200}$ of an inch each in diameter. They are formed of a delicate layer of mucous membrane, closely attached to which are minute plexuses from the pulmonary artery and veins, and to unite the whole there is a quantity of yellow elastic tissue.

According to the calculation of M. Rocheaux, there are 17,790 air-cells connected with each terminal bronchus, and in the lungs, 600,000,000. Prof. Dalton, of this city, estimates the amount of surface thus exposed to the action of the air, at 1,400 square feet. The capillary vessels of the pulmonary artery and pulmonary veins distributed in delicate meshes on the walls of the air-cells are the channels through which the blood-changes are effected. The venous blood, loaded with carbonic acid, is carried by the pulmonary arteries from the right side of the heart to the lungs, where it gives up its load of impurity. The capillaries of the pulmonary veins, which originate in the walls of the air-cells, take up the renovated blood with its load of oxygen, and carry it to the left side of the heart, whence it is propelled to all parts of the body.

The interchange of gases and aëration of the blood are effected during the respiratory movements of inspiration and expiration. During inspiration, the diaphragm contracts and increases the vertical diameter of the chest, while the ribs are elevated and separated by the action of the other inspiratory muscles, thereby making the lateral diameters greater. A vacuum is thus formed, and the air rushes in. Following immediately is an expiratory movement, in which the air is forced out: 1. By the relaxation of the diaphragm, which is pushed upward by the abdominal organs resuming their original positions; 2. The ribs are drawn together by the external intercostals; and 3. The lungs, which are extremely elastic, contract and force the air out of the cells. After the air enters the bronchial tubes, a diffusion of gases takes place, and the impure air below passes upward, while the oxygen continues on to the air-cells. After reaching the cells, the oxygen passes by endosmosis through to the blood, and is carried off by the corpuscular elements of the circulatory fluid which have previously given up their carbonic acid. Allowing that twenty respiratory movements take place in a minute, the air in the lungs will be necessarily changed 1,200 times in the course of an hour. About 17 cubic feet of oxygen are consumed in 24 hours, and during the same period from 300 to 400 cubic feet of atmospheric air are supplied to the lungs.

Oxygen gas is an essential requirement of a healthy organism. It exerts a remarkable influence upon both vegetable and animal life. Eight-ninths of the whole mass of water, one-third of the earth's substance, and one-fifth of the atmosphere, are composed of oxygen: no element is more abundant or more important.

Repair and decay are closely linked in the animal economy. Death is a necessary accompaniment of life. Molecular disorganization, elaboration, and growth of new material, proceed simultaneously. In health the growth keeps pace with loss, in disease waste preponderates. During the physiological interchange of material new substances of a poisonous nature are generated, and are removed by the different emunctories. Should the avenues of escape be closed, life is speedily terminated. For instance, the kidneys eliminate an excrementitious substance called *urea*, which is formed by the decay of nitrogenized tissue. When these organs cease to abstract this material from the blood, it accumulates and produces convulsions, coma, and finally death. Carbonic acid, which is specially under consideration in this connection, is another product of retrograde metamorphosis. When through disease or accident it is retained, and the blood imperfectly aërated, all the nutritive processes are retarded, or entirely stopped. The vitality of the body necessarily fails to approximate to a healthy standard, and latent germs of disease are impelled to an inordinate and even fatal growth.

The large mortality in our tenement-houses is sufficient evidence of this truth. Human beings are crowded together in these dens, in a stifling atmosphere, unfit to supply the wants of the system. A family of six and seven will sometimes be cramped in one or two small rooms, scarcely large enough to accommodate a single person. But it is not alone the evil of a diminished quantity of oxygen that these people have to contend with; the surrounding atmosphere is rendered doubly poisonous by the animal exhalations which naturally accumulate and occasion cholera, typhus fever, and other pestilences. In these homes of the poor, these monuments to the grasping spirit of the nineteenth century, death reaps a rich and continuous harvest. And all this must endure until the strong arm of the law compels avaricious landlords to construct houses properly ventilated, and fit for human habitation.

As an example of the effects of imperfect ventilation, the suffocation of a large number of persons in the famous or rather infamous " Black Hole of Calcutta," will be remembered. One hundred and fifty persons were confined for a single night in a room eighteen feet square, having but one small window. In the morning only seventeen were alive. As another example of the evils attending imperfect ventilation, we may mention the destruction of life which occurred on an Irish steamer some years ago while crossing the Channel. During a storm the captain compelled one hundred and fifty of the passengers to go below, and afterward closely fastened down the hatchways. Seventy persons perished before the hatchways were removed. The violent storm prevented their outcries from being heard, otherwise their horrible fate might have been averted.

Similar occurrences, but on a smaller scale, are frequently brought to our notice. They generally arise from design or neglect.

The condition resulting from a complete cessation of the respiratory movements is usually known as *asphyxia* or *apnœa*. The word *asphyxia*, derived from two Greek words signifying pulselessness, does not define the condition. *Apnœa* indicates the prominent features of the morbid process with greater accuracy; but, as asphyxia is the word in general use, it will be adhered to in the present chapter.

The first effect of obstructing the entrance of air is a retardation of the current of blood in the capillary vessels of the lungs and general system. The blood accumulates and moves slowly through them. Should the ingress of air be still further prevented, this state of congestion ends in complete stagnation or stoppage of the circulation. Unaërated blood cannot pass through the capillaries.

Prof. Austin Flint, Jr., considers the want of oxygen in the tissues, and the accompanying capillary congestion, as the starting-point of suffocation or *asphyxia ;* and that the obstruction in the capillaries throws the blood

back on the heart, and overpowers it, so that it entirely
ceases.

Some consider that the congestion of the lungs is alone
the cause of death; others, that the blood going to the
brain, loaded with carbonic acid, destroys the activity of the
cerebrum, and through it acts upon the heart and the nerves
supplying that organ.

Where so many phenomena exist, involving different
vital parts, it is almost impossible to separate them, and
definitely say which is the cause of death. To repeat, defective aëration causes the rapid increase of carbonic acid, and
induces capillary congestion in every part of the system; this
congestion demands more labor from the heart, and the congestion of the lungs increases the difficult respiration, and
makes it more labored. The blood, which is loaded with
carbonic acid, necessarily obtunds nervous sensibility, and,
acting through the cardiac nerves upon the heart, combines
with the other morbid influences in weakening the contractions of that organ, and bringing about a fatal termination.

The morbid appearances after death vary but little with
the cause of the asphyxia. In the majority of cases there
is a similarity in the changes. The face generally is of a
dark, livid color; froth or foam, streaked with blood, sur-
rounds the mouth. The eyes protrude. In suffocation from
hanging, the tongue is swollen and pushed out between the
lips  Rigor mortis appears soon after death. The lungs
are heavy and dark, and contain a large quantity of black
blood. The air-cells and smaller bronchial tubes are filled
with a sanious, frothy fluid. Blood is absent from the left
side of the heart and arteries. This latter peculiarity is due
to the elasticity of the walls of the arteries forcing out the

blood. It is not confined especially to death from suffocation, but occurs in other forms.

The auricle and ventricle on the left side of the heart are distended with dark blood, and all the blood in the body is blacker than under ordinary circumstances. This is caused by the absence of oxygen, which gives the circulating fluid a red color. In the liver, kidneys, and spleen, there is generally more or less congestion. There are various opinions advanced respecting the conditions of the brain. Some modern investigators (*Ackerman*, *Dondus*) endeavored to show that anæmia of the brain is more common than congestion. This idea, however, is not sustained by facts, or accepted by many in the profession. The cerebral vessels, except in rare cases, are engorged with blood.

Having now dwelt on the physiology of respiration, and the pathological changes which depend upon the defective aëration of the blood and total cessation of the respiratory act, we now come to the various forms of asphyxia and their treatment.

STRANGULATION.—This term is generally applied to that variety of asphyxia caused by external compression; but any mechanical occlusion of the trachea or larynx, whether external or internal, belongs under the same head.

The strangulation produced by clasping the throat tightly with the arm or hands is the common method employed by garroters. In suicidal attempts, handkerchiefs or ropes are generally used, and the rope is resorted to in most civilized countries in judicial strangulation. All cases of hanging, however, do not terminate by asphyxia. The neck is usually broken by the fall, and death results from pressure

10

on the upper part of the spinal cord, and congestion of the brain.

The greatest number of strangulated patients who come under the care of the surgeon are those of attempted suicides, and every stage of asphyxia, from a slight suffocation to complete stoppage of respiration, may be found among them.

The symptoms arising from mechanical occlusion of the air-passages are common in a greater or less degree to all other varieties of asphyxia. They are usually so well marked as to preclude a possibility of mistake. At the same time, the history of the patient should always be inquired into. The patient's countenance presents an anxious expression, and is of a livid color, which, in extreme cases, is almost black. The lips are swollen and somewhat everted, the eyes bloodshot and prominent, the vessels of the head and neck are enlarged to double their ordinary size. There is an intolerable feeling of discomfort and oppression over the chest, and intense desire for air. The respiratory movements become rapid, but after a time they are slow and prolonged. There is a momentary increase in the pulsations of the heart. As the asphyxia progresses, the movements diminish in force, until they are totally lost. In the beginning, the patient suffers from giddiness, ringing in the ear (*tinnitus aurium*), and great general distress. The agony gives way where asphyxia results from immersion in water, and is succeeded by pleasant visions and dreams. In some recorded cases, these sensations are said to have been so entrancing as to cause the resuscitated patient to curse his attendants for bringing him back to renewed torture. These dreams are followed by insensibility; the pulse is usually

absent, but the action of the heart may still be made out
with a stethoscope. So long as an impulse is detected,
there is chance of recovery.

In asphyxia resulting from violence, there is often an accompanying condition of syncope. This may resemble death
to such an extent as to prevent the continuance of treatment.
However, if the points of difference between death and simple
insensibility are appreciated, there will be little difficulty.

When life ceases, the pupils are dilated, the cornea is
flattened, and the eyes fixed. There are congestion of the
cutaneous capillaries, especially in the most dependent portions of the body, and blueness under the finger-nails. (In
true asphyxia this congestion is not a sign of much importance.) All respiratory movements have ceased, and no
moisture will appear on a looking-glass held over the
mouth or nose. The pulsations of the heart cannot be
made out with the ear or stethoscope.

Another test has been proposed lately by a French gentleman, who states that, if a bright steel needle be inserted
into the dead body, it will become tarnished; if introduced
into the living body, it will come out perfectly clean. If a
preparation of Calabar bean is applied to the eye while life
is present, the pupil will contract; if death has taken place,
no effect will be produced.

A muscular rigidity (*rigor mortis*) ensues soon after
death, and a peculiar, offensive odor is emitted.

*Treatment.*—The treatment of strangulation from foreign bodies in the air-passages has been considered in a previous chapter. When strangulation results from external
compression of the throat, a careful examination should be
instituted to ascertain the amount of local injury. Lacera-

tion or fracture of the larynx or rings of the trachea may cause pieces of cartilage to protrude on the internal surface. Such obstructions must be removed, in order to render the treatment effectual. All superfluous clothing should be removed from the chest and neck, and the mouth and throat cleared of mucus. Artificial respiration, either by Marshall's, Hall's, or Sylvester's methods, must then be tried. The manner of employing these methods is hereafter fully explained. It is at times necessary to perform tracheotomy (*see* Tracheotomy), and to fill the lungs by forcing in air with a bellows, or with the mouth applied to the opening. In addition to artificial respiration, the surface of the body should be briskly rubbed to keep up the circulation, and stimulants administered through the rectum. As in cases of hanging there is congestion of the brain, a few ounces of blood can be taken from the arm with benefit.

COMPRESSION OF THE THORACIC WALLS produces suffocation by preventing the expansion of the lungs and admission of air. It usually occurs from jamming, or by being crushed beneath embankments or masses of building material. In the former case the sufferer is usually very much frightened. The arms are thrown involuntarily above the head, leaving the chest exposed to the pressure of the crowd. Persons in large crowds can, with ordinary precautions, protect the chest by keeping the arms and elbows close to the side of the chest, flexing the forearm, and bringing it in front, thus making the hands meet in the median line. Unless extraordinary pressure is made, this method will allow of sufficient respiratory movement to sustain life.

The notorious resurrectionist and murderer, Burke, usually destroyed his victims by compressing the thoracic walls.

With this variety of asphyxia there may be more or less bruising and laceration of the chest-walls, but the general symptoms and treatment are the same as given above.

SUFFOCATION FROM INHALATION OF GASES.—The inhalation of nitrogen or hydrogen occasions the same changes and symptoms as are witnessed in other forms of asphyxia. Nitrogen exists in large quantities in atmospheric air. When inhaled in a pure state, it destroys life with greater rapidity than other gaseous bodies.

The inhalation of sulphuretted hydrogen, carbonic acid, carbonic oxide, carburetted hydrogen, etc., should be treated under the head of poisons. As death in these cases, however, is usually attributed to asphyxia, and as the treatment is the same, they will be considered in this section.

*Sulphuretted hydrogen* is a product of the decomposition of animal matter. It is found in sewers, old drains, and stagnant pools. The foul odor of "rotten eggs" is due to this gas. When inhaled, it proves rapidly fatal. According to Flenard, one part in a hundred and fifty of atmospheric air will kill a horse. Men can bear larger proportions.

Small quantities of sulphuretted hydrogen, inhaled in a diluted form, give rise to nausea, vomiting, pains in the abdomen and extremities, vertigo, and a semi-paralytic condition of the extremities. In large quantities, it produces rapid insensibility, convulsions, and death. The body exhales the characteristic odor of the gas. After death the mouth and fauces are coated with a dark-brown mucus. The muscles and all the internal organs are dark-colored, and the blood is fluid.

*Carbonic acid*, or di-oxide of carbon, is found in large quantities in the bottom of wells, coal-mines, and in all dark, damp situations, where organic matter is in a state of decomposition. In coal-mines it is usually known as "choke-damp," and death is often caused by its inhalation. This substance also results from the physiological decay of living bodies.

An atmosphere containing one-tenth of carbonic acid will produce death. Its effect on the system is that of a narcotic poison, although, when death results from its in-halations, it is commonly said to cause suffocation, or asphyxia.

The symptoms attending its inhalation, with one or two exceptions, resemble those occurring in ordinary asphyxia. There is at first marked loss of muscular power, with tendency to sleep, and the countenance assumes a leaden hue. After death the eyes remain bright for some time, and several hours elapse before rigor mortis sets in.

*Charcoal-vapor* consists of carbonic acid, carburetted hydrogen, free nitrogen, and atmospheric air. This vapor is often used as a means of self-murder. In France it is frequently employed for this purpose. Suicides burn the charcoal on a brazier, in a close room, where all the crevices for the admission of air are shut off. The vapor at first creates a sensation of extreme languor and general weakness. This is soon followed by complete insensibility. In some of these cases the countenance is pale, and the jaws are usually fixed. After death the heart is empty, or a little black blood may occupy its right ventricle.

*Coal-vapor.*—The materials arising from the ordinary combustion of coal are sulphurous acid, carbonic acid, sulphuretted hydrogen, and carburetted hydrogen. It is impossible to inhale this vapor under ordinary circumstances. It possesses such irritating qualities that, unless a person is stupefied with alcohol or other narcotics, he will escape before a sufficient amount is taken in to destroy life. Occasionally, persons are suffocated in holds or cabins of vessels from this vapor. A sad instance occurred recently in New-York harbor. Five seamen shut themselves in the forecastle, where a brazier of coal was burning, and in the morning were found dead.

*Coal-gas.*—This substance is employed for illuminating purposes. It consists principally of light carburetted hydrogen, carbonic oxide, olefiant gas, hydrogen, nitrogen, etc. Its odorous principle is due to vapor of naphtha. *Carbonic oxide* is supposed to be its principal poisonous ingredient.

If the atmosphere of a room becomes impregnated with twelve per cent. of the gas, a lighted candle introduced will cause an explosion. Accidents arising from coal-gas are generally the result of carelessness or ignorance. Neglecting to turn the gas off, and leakage in the pipes, are the common causes. The effects produced by its inhalation differ from other varieties. There are more or less vertigo, nausea, and vomiting, a semi-paralytic condition of the muscles, and convulsions ending often in death. After death the blood is sometimes of a light-red color.

*Treatment.*—In all these varieties of suffocation, inhalation of oxygen gas will bring about speedy relief. Where respiration has ceased, it must be restored by artificial methods. If necessary, oxygen may be forced into the lungs in the manner previously mentioned. Cold water, poured on the surface of the body, is likewise beneficial.

### DROWNING.

The length of time that persons can remain under water, and afterward be resuscitated, varies according to the circumstances attending each individual case. When timid persons become accidentally submerged, they throw up the arms, open the mouth to shriek, and consequently fill the lungs with water and strangulate at once. If presence of mind is not lost, the arms kept under water, and the respiratory movements controlled until the head comes above the surface, life may be prolonged a considerable period. Again, should the submerged individual faint, the chances of resuscitation are good even when several minutes have been spent without air. The fit of syncope is attended with a stoppage of respiration and of the heart's action, and, the demand for oxygen being diminished, the system does not feel the loss as it would under other circumstances. Occasionally, life is destroyed after an immersion of one minute, while in other instances persons remain under water for two and even three minutes without receiving injury. Thus sponge and pearl divers, who spend a great part of their working-hours under water, remain deprived of air for two or three minutes with but little discomfort. Marac relates the case of a German woman who was tied up in a bag with a cock and cat, and thrown into the water as a punishment for child-murder. She was submerged fifteen minutes, and, when removed from the bag and exposed to the air, immediately recovered. Such a prolongation of life without air can only be accounted for on the supposition that the woman fainted on being immersed, and that the state of syncope lasted until she was brought to the surface.

A committee of the Royal Chirurgical Society, London, instituted a series of experiments to ascertain the length of time animals could sustain life without a supply of oxygen. A brief statement of the principal results will be of interest. It was ascertained that, when the entrance of air was prevented by submersion, death was more rapid than when the trachea was thoroughly closed with a plug. When the trachea was simply plugged, the respiratory movements continued from three to four minutes and a half, and the action of the heart was perceptible from six to seven minutes and a half. As a rule, the heart's action continued two or three minutes after respiration ceased. When animals were kept under water one minute and thirty seconds, death followed, even when the animal was taken out alive. No efforts were made in any of these cases to restore life. If respiration had been artificially produced, they would have probably recovered. The striking difference in the period of death in the two classes is explained by the fact that, in simple plugging of the trachea, sufficient air remained in the lungs to maintain life for a short time, while in the other, water found its way into the lungs and displaced the air which might otherwise have been reserved for aëration. Some contend that water does not enter the lungs of the drowned, but the results of *post-mortem* examinations do not confirm this statement. Water, sea-weed, and other extraneous matter, have been found in the bronchial tubes in the majority of cases. It is true that at times there is not the slightest trace of water. This circumstance is, however, exceptional. The remarkable power of absorption possessed by the lungs may account for the rapid disappearance of the liquid. As human beings, when drowning, alternately sink below, and rise again to the surface of the water, occasionally giving them opportunity to obtain a fresh supply of air, we cannot definitely determine the maximum of time they can remain under water and yet recover afterward. The experiments quoted are not proper criteria to judge by in the majority of drowning cases. When submersion is continuous, however, five minutes is the longest period after which life may be restored. There is a peculiar condition, known as secondary asphyxia, which occurs at times in persons who have been restored by artificial respiration. It shows itself generally within forty-eight hours after respiration has been fully established. When the symptoms seem favorable, and all anxiety removed, the patient is suddenly seized with urgent dyspnœa, the chest expands imperfectly and irregularly, the patient struggles for breath, and in a short time all the worst features of asphyxia return. Death soon supervenes, unless immediate relief is afforded by artificial respiration. The cause of this change is not well understood. It is probably due to congestion of the lungs, induced by some active movements on the part of the patient. The exercise sends more blood to these organs than they, in their weakened condition, can provide for. Excessive and laborious respiration immediately follows. The appearances presented in asphyxia resulting from immersion vary somewhat from other kinds. The livid discoloration of the face and fulness of the blood-vessels are not so distinctly marked. There are more general pallor and coldness of the surface. Rigor mortis or *post-mortem* contractions of the muscles appear very soon after death.

*Treatment.*—There are four special requisites in the treatment of drowned persons : 1. *Artificial respiration ;* 2 *Warmth ;* 3. *Friction ;* 4. *Stimulation.* All these are employed together, but the first is generally relied on. Strip the patient of clothing, and envelop the body as far as possible in warm blankets. Then clear the mouth and throat of water, mucus, or other substance which might prevent the ingress of air.*

To do this perfectly, cover the index-finger closely with a handkerchief, and carry it in as far as possible, and sweep it around the pharynx and upper part of the larynx. The cloth takes up more of the moisture than the finger alone would. The tongue is now drawn out as far as possible. Unless the organ is pulled forward with considerable degree of force, the aryteno-epiglottidean folds at the upper border of the larynx will close the aperture sufficiently to interfere with the admission of air. This is a point of considerable importance in all cases where artificial respiration is resorted to, and cannot be too strongly insisted upon. A forceps attached to the extremity of the tongue, or a towel wrapped around its end, and grasped with the thumb and forefinger, will make traction easy. Having cleansed the air-passages, we try some of the methods of artificial respiration. When the immersion has been short, and the patient only partially asphyxiated, simple compression of the lower half of the thorax and upper part of the abdomen will answer. The hands are applied on each side of the chest-walls, the fingers reaching as high as the nipple, and firm

---

* Some advise suspension of the drowned person by the limbs, in order to facilitate the escape of water from the lungs ; but this is an unnecessary pro cedure.

pressure made to diminish the cavity of the chest. The hands are then lifted for a few seconds, and the parts allowed to resume their natural position.

This is done rapidly and continuously until all danger has passed. Diminishing the thoracic cavity by pressure forces out some of the foul air from the lungs, and with the subsequent expansion a certain amount of fresh air passes in. This interchange gives more oxygen to the blood, and relieves it of carbonic acid, stimulates the circulation, and through it the nervo-muscular apparatus, and finally restores all the functions of life. In severe cases, either Marshall Hall's or Sylvester's method of artificial respiration is to be preferred. The latter is said to be superior, as it enables more air to pass out of and enter the chest. The preliminary steps, such as clearing the throat and drawing out the tongue, are the same. In Marshall Hall's method the patient is placed on the side, with the arm toward the posterior plane of the body. The body is then rolled slowly over on the face, while the hands of the surgeon at the same time are pressed firmly on the back and sides of the chest, diminishing its cavity. When this movement is completed the patient is turned on his back, and the chest-walls resume their original position; these movements are to be kept up until natural respiration is resumed. The principal effect to be produced in all cases is a renewal of the air in the lungs. In Sylvester's method the patient is placed in the recumbent position, with the head and chest somewhat raised. The operator stands at the head of the patient and grasps both arms midway between the elbow and wrist-joint, moving them gradually to a vertical position so as to make them nearly meet above the head. They are held in this position for a moment, and then slowly returned to the sides. At the termination of the second movement, pressure is made with the arms on the sides of the thoracic walls. These movements are continued as long as the asphyxia remains. Raising the arms in this manner elevates the ribs, and allows comparatively a large quantity of air to enter, while relaxation causes them to resume their normal relations. Conjointly with all varieties of artificial respiration, the patient's limbs should be briskly rubbed by an assistant, and brandy and ammonia should be administered through the mouth or rectum. Hot bottles and blankets are to be applied to the extremities before and after the patient has recovered. Heat, by means of hot-air baths, is sometimes useful. Ammonia, in the form of vapor or in solution, may be applied to the nostrils. Should ordinary artificial respiration fail to revive the patient, pure oxygen may be forced into the lungs. This may be done by cutting a hole in the trachea, inserting a tube, and forcing the gas through it. The ordinary elastic bag employed for inhalation of oxygen, if pressed with moderate force, will send in enough gas to distend the lungs. If the gas is not at hand, the nozzle of a bellows may be attached to the trachea-tube, and the necessary expansion accomplished with atmospheric air.

Injuries to the Spinal Cord, above the origin of the phrenic and intercostal nerves, paralyze the muscles of respiration and produce death by asphyxia. Poisonous doses of nux-vomica or its alkaloids cause spasm of the same set of muscles, and terminate life in like manner.

# CHAPTER XIII.

*SUNSTROKE.*

*Synonymes.*—Heat Apoplexy—Insolation—Sun-Fever.

We have records of sunstroke from the earliest histori-cal times. It is fully described by ancient medical writers. About the first cases mentioned are the following, from biblical history:

" Manassas was her husband, who died in the early har vest: for, as he stood among them and bound sheaves in the field, the heat came upon his head, and he fell on his bed, and died in the city of Bethuliah." The second in-stance relates to the son of the Shunammite woman, who was restored to life by the prophet Elisha: " And when the child was grown, it fell on a day that he went out with his father to the reapers. And he said unto his father, '*My head, my head.*' And when he had taken him and brought him to his mother, he sat on her knees till noon, and then died."

Sunstroke is not confined to tropical regions; New York and other Northern cities suffer from its yearly visitations. At certain seasons the number of cases, in proportion to the population, far exceeds that of the more tropical towns. In New York, especially, the mortality has been very great.

During the summers of 1866 and 1868 an immense number of cases were recorded.

Visitors to the tropics from the colder regions, who are unaccustomed to a high temperature, are particularly susceptible; while the natives, who live constantly exposed to the heat, are comparatively safe.

Sunstroke does not depend upon a short exposure to the direct rays of the sun; the exposure must have been continued for a day or two; nor does it necessarily arise from solar heat. Prolonged confinement in the heated atmosphere of a building may likewise produce it.

Dr. Maclean * speaks of thirteen cases which occurred under Mr. Longmore, in the barracks at Burrackpoor, India, while only three arose from outside exposure. The same thing has been witnessed on crowded vessels, in laundries, and sugar-refineries. I recall three fatal cases which were admitted to Bellevue, of persons who were prostrated while at work in a sugar-refinery. Dr. Swift gives the history of twelve persons who, while at work in a large laundry in this city, were similarly affected. Some of these patients may have been exposed to the solar rays, but the majority were at work in-doors.

About the third or fourth day from the commencement of a heated term, sunstrokes usually appear. The sufferers in most cases are exposed to the heat for some days preceding the attack.

In the summer of 1866 the majority of sunstroke cases —generally laboring-men—were brought to Bellevue Hospital in the morning or early in the day.

* Reynolds's Practice, article Sunstroke, p. 156.

Persons of intemperate habits and debilitated systems are most liable to attack. Any thing tending to lower the vitality of the system predisposes to the affection. Wearing heavy, dark clothing, or compressing the chest, is also unsafe. The close-fitting regulation uniform and equipments worn by the British soldiers in India swelled the bills of mortality from sunstrokes when that country was first occupied. Better sanitary ideas of soldiers' dress have been developed within the past few years, and the death-list has consequently diminished.

Sunstrokes may be classed under two heads: 1. Those in which the nerve-centres are principally involved, or the cerebro-spinal variety of Morehead; 2. The varieties which are characterized by exhaustion. Death in the former case results from *coma;* in the latter, from *syncope.* In some forms death is ascribed to asphyxia, or apnœa.

Persons of full habit addicted to the use of spirituous liquors are generally victims of the cerebro-spinal variety. Hard-working individuals are more liable to the cardiac form.

In typical cases of sunstroke the symptoms may be divided into premonitory and immediate. The premonitory symptoms are not always evident. The patient complains of headache and a burning sensation about the head, and during the night is restless and wakeful. The skin is dry and uncomfortably hot, and there is frequent desire to evacuate the bladder. The face is flushed, and eyes congested; the bowels are usually constipated. A person presenting these symptoms, who, nevertheless, continues to work under the hot sun, or in an overheated building, will be suddenly seized with vertigo, intense headache, and dimness of vision. His limbs refuse to support him, and he soon falls to the ground. Insensibility sets in; the breathing becomes stertorous, pupils contract, and the skin is intensely hot. The temperature of the body, ascertained by a thermometer in the axilla, varies from 100 to 107, in rare cases reaching 109. The pulse is rapid, and often full; as the case progresses toward a final termination, it becomes weaker and irregular, but still very rapid. The coma may be either partial or complete, and occasionally there are convulsions. The bowels are sometimes relaxed, and vomiting is not infrequent.

There are various grades or manifestations of sunstroke. Some who come under the physician's care complain of intense weakness, and pain in the head. Others are stupid and wandering, while complete insensibility accompanies the great majority of cases. In some the general *malaise* and warning symptoms precede the insensibility for several days; others are stricken down in a moment, without previous uncomfortable sensations.

In those varieties of sunstroke characterized by exhaustion or syncope the patients are more apt to die suddenly without special premonitory troubles. In such cases the countenance is paler than in the cerebro-spinal variety. The respiration is sighing or gasping instead of being stertorous. The pulse is generally rapid, compressible, and irregular. The pupils may be dilated, the heat of the skin is not extreme; sometimes there is a combination of the cardiac and cerebro-spinal varieties.

The reason why consciousness is lost, from exposure to extreme heat, is not fully understood; overheating of the blood is said by some authorities to call for excessive action

11

in the nerve-centres, which rapidly exhaust their force and power.

Maclean and others regard the heated blood as producing great depression of the nervous system, and thus preventing it from performing its functions. The latter theory seems the most plausible.

Even if we accept this view, there are changes in the nerve-fibres and cells which we have as yet been unable to recognize or fully understand. These changes, in many cases, make recovery from sunstroke more to be dreaded even than death itself. They give rise to the varied sequelæ of sunstroke, such as amaurosis, obstinate and distressing headache, and impairment of the intellect.

Insanity in its varied forms is a common sequence. In some instances, the brain is found to be softened after death, in others there is no special lesion perceptible.

On *post-mortem* examination the brain and its membranes are usually found to be congested. In persons who die from exhaustion this feature is less marked. The great mass of cases, however, show this change. Out of twenty-two *post-mortems* which I made in Bellevue, twenty presented cerebral congestion. All had marked congestion of the lungs. Two of them showed evidences of inflammation in the mucous membrane lining the stomach and intestines. Before death they had violent attacks of vomiting and purging. Congestion of the lungs is almost always present. The right side of the heart is distended with blood which is entirely fluid, and without tendency to coagulate. Decomposition proceeds rapidly after death from sunstroke.

*Treatment.*—It was considered imperative at one time to abstract blood in all cases of sunstroke. Modern enlightenment has excluded this therapeutical agent. Depleting measures of every kind are now considered injurious.

The patient should be removed at once to a cool room, and placed in a recumbent position near an open window. The clothes are then stripped off, and a stream of water poured over the body. The vessel containing the liquid is to be held about four or five feet above the patient, in order that he may receive the benefit of the shock. The stream of water should at first be directed on the head, then on the chest and abdomen, and finally on the extremities, and thus alternating from one part to another, until consciousness returns. Ice rubbed over the body is liked by some; the cold douche is, however, preferable.

When the dyspnœa is marked, a few dry cups placed on the thorax in front and behind will be of service.

Internal medication is useful in all cases. Among the numerous drugs employed, bromide of potassium has been found most efficient. The best results were obtained from its use in Bellevue Hospital, in the years 1866 and 1868. This drug may be administered in all stages of the affection. When the patient is unable to swallow, it can be given by injection, always remembering to increase the dose one-quarter more than when given by mouth. In mild cases from five to ten grains may be given, at intervals of from half an hour to one hour, until the grave symptoms disappear. In several forms from ten to thirty grains may be administered every half-hour; when the pulse becomes weak or intermittent, stimulants are needed. Stimulation should be resorted to in all cases where exhaustion is the prominent feature. Brandy-and-milk, or brandy with ammonia, must be introduced into the stomach or rectum.

The cold douche must be sparingly employed, or altogether dispensed with in this latter class of cases. If the skin is cold, it will do no good whatever.

After consciousness has returned, mustard-plasters or blisters are to be applied to the back of the neck or behind the ears. The bromide need not be discontinued for one or two weeks.

As soon as convenient, the patient should be sent to a cool district in the country, and kept free from all sources of excitement. The brain must rest from all work. Exercise in the open air and nourishing diet are essential; regular habits must be rigidly enforced. A continuance of this treatment for several months prevents or at least lessens the danger from nervous affections which follow sunstroke.

# CHAPTER XIV.

## *DYSPNŒA.*

Dyspnœa from Asthma—Croup—Congestion of the Lungs—Cardiac Disease
—Pulmonary Œdema—Pulmonary Apoplexy, etc.

SHORTNESS of breath or difficult respiration arises from defective aëration of the blood. Any condition which diminishes the amount of oxygen sent to the tissues, or creates a demand for more than the lungs in ordinary respiration can furnish, will occasion dyspnœa. Over-exertion produces the simplest illustration of the manner of its production. Violent muscular movements quicken the cardiac impulses, and a larger amount of blood is sent to the lungs as well as to other organs. There follows a demand for more oxygen, and the respiratory movements are increased to make up by rapidity of inhalation the diminished quantity of that element in the blood.

In the category of diseases characterized by dyspnœa are included asthma, croup, congestion of the lungs, cardiac affections, pneumonia, bronchitis, pulmonary œdema, pulmonary apoplexy, and œdema glottidis. The dyspnœa which is caused by mechanical obstruction or occlusion of the air-passages is considered in another chapter.

ASTHMA.—In this disease there is a spasmodic contraction of the muscular fibres of the smaller bronchial tubes, and a consequent diminished calibre of these tubes, which prevents the free ingress of air. Asthma exhibits a preference for certain localities and seasons of the year. It may occur at any season, but prevails specially in the autumn. It is said to be caused in some instances by the inhalation of new-mown hay, ipecac, coal-dust, and other substances. Inflammation of the bronchial tubes also excites it. It is not an unfrequent accompaniment of emphysema.

The paroxysms usually develop suddenly. The patient struggles for breath, and runs to the open window. The respirations are not quickened. A wheezing noise is heard with each respiratory movement. The voice is low and husky. The face is congested, the lips blue, and the eyes prominent. A cold perspiration appears on the surface. The pulse is small, and in some cases very rapid. There is inability to maintain the recumbent position. The patient usually sits bent forward and resting on his knees, bringing every auxiliary muscle of respiration into use to obtain air. On auscultation, loud sibilant and sonorous râles are heard over both lungs. The attack usually lasts from half an hour to four or five hours; but it may continue with varying degrees of severity for two or three days.

The absence of œdema, valvular lesions, febrile excitement, etc., and the comparative good health between the paroxysms, are sufficient to distinguish the disease.

*Treatment.*—Pure oxygen has lately been employed with considerable benefit in this disease. Five or six gallons should be inhaled every fifteen or twenty minutes until relief is experienced. Even where it does not completely subdue the paroxysm, it will at least diminish the distress.

Chloroform, ether, and other anæsthetics, may also bo given with advantage. There are some cases which can only be relieved by these medicines.

The majority of practitioners employ simple antispasmodics, such as stramonium, belladonna, or lobelia. The former drug may be given in two-grain doses every half-hour, or the leaves may be smoked in a pipe, or in the form of cigarettes, until relief is obtained. Hoffman's anodyne may be used in conjunction with inhalation of steam. A basin of hot water is held under the patient's head, the anodyne is poured slowly in, and the ethereal vapor mixes with the steam, and is inhaled. A blanket thrown over the head of the patient prevents the steam from escaping. Belladonna in quarter-grain doses of the extract relieves certain varieties of asthma with great rapidity. Emetic doses of lobelia, eupatorium, or ipecac., are recommended by some.

CROUP.—There are two principal varieties of this disease, viz., membranous and spasmodic. The first is an inflammatory affection, attended with fibrinous exudation, and is usually fatal. In the second there is a spasmodic contraction of the muscles which govern the vocal cords. It may appear with or without catarrh of the larynx, and is rarely if ever fatal. As the spasmodic variety is more rapidly developed, and as a rule unattended by premonitory symptoms, it may properly be considered a case of emergency, and discussed in this connection.

The spasm of the vocal cords which occurs in spasmodic croup may arise from the reflex irritation of worms in the alimentary canal, from teething, or from a cold or catarrh. The attack comes on in the night. The child wakes from its sleep with a loud, heavy, croupous cough, husky voice, and intense dyspnœa. The face becomes dusky and livid, and the extremities are cold. In a short time the spasm relaxes, and the child resumes its natural breathing; but the hard cough and changed voice remain longer. If the attack be connected with catarrh, the hoarseness is more likely to continue, and the paroxysms will recur at various intervals during the night. It is differentiated from membranous croup by the absence of exudation on the tonsils, constitutional and local signs of inflammation, and also by the fact that in spasmodic croup there is complete relief between the paroxysms. In the membranous or true croup the dyspnœa continues or increases as the disease advances.

*Treatment.*—An emetic composed of a drachm or two of the wine of ipecac., or four or five grains of the powder, should be administered without delay. The child should then be immersed in a hot bath for five or ten minutes. When taken out, warm blankets should be wrapped around the body, and hot flannels or hot hop-poultices applied to the throat. To prevent a recurrence of the paroxysm, all sources of irritation should be removed, and the general health sustained by attention to diet, nutritious food, good air and exercise. If there be a strong predisposition to these attacks, small doses of bromide of potassium, belladonna, valerian, etc., may be given with salutary effect.

Membranous croup is treated by inhalation of steam, oxygen, and internal administration of iodide of potassium; tracheotomy is sometimes performed. Recovery is rare.

CONGESTION OF THE LUNGS.—DYSPNŒA which occurs from engorgement of the pulmonary capillaries is rarely as sudden in its origin as that which arises from croup or asthma. Congestion is due to a variety of causes. It is an accompaniment of pneumonia and bronchitis, and is a fatal element in the latter stages of cardiac disease. Patients with valvular lesions or other organic affection of the heart are after unusual exertion liable to congestion. The debilitated heart beats with greater rapidity and violence, and the lungs, already overloaded with blood, become rapidly engorged. The respiratory movements are almost doubled in endeavoring to introduce the necessary amount of oxygen.

The patient sits up in bed, moving the head from side to side, and gasping for breath. There is an expression of great anxiety, and the face is bathed in cold perspiration, and marked by the characteristic cyanosis. The pulse is irregular, rapid, and intermittent. Sometimes the overloaded blood-vessels relieve themselves by rupture, and pour out blood into the parenchyma of the lung, and into the bronchial tubes. If the extravasation is great, a fatal termination is reached in a short time ; a small hæmorrhage is of little consequence.

*Treatment.*—Medicines which diminish the frequency of the heart's action are indispensable. Digitalis is the best remedy we possess for the purpose. Aconite and veratrum viride are preferred by some. Digitalis may be given in powder, tincture, or extract. The tincture is the most reliable preparation. It may be given in five-drop doses every half-hour until the patient is relieved. With the internal medication the application of a dozen dry cups to the chest is called for. If the patient is not very much debilitated, a few wet cups may be applied. Inhalations of oxygen gas are also beneficial. The subsequent treatment consists in restraining the patient from all active exercise, and keeping the action of the heart within proper limits. Every source of mental excitement must be avoided. Tonics, good diet, and fresh air, are always necessary.

Congestion dependent upon pneumonia or bronchitis is relieved by cathartics, counter-irritation by means of blisters, abstraction of blood with wet cups, and promoting diaphoresis by small doses of antimony or ipecac.

PULMONARY ŒDEMA is induced by conditions which give rise to œdema in other parts of the body. It occurs in cardiac disease, and in degeneration of the kidneys. The serum is poured out from the distended vessels into the air-cells and areolar tissue of the lungs. Both lungs usually are affected. In the recumbent position the serum gravitates to the posterior portion of these organs. The exudation usually takes place gradually, but it may be poured out so rapidly as to destroy life in a few moments.

Urgent dyspnœa marks its occurrence. The patient's face and limbs may be swollen from œdema, or other signs of Bright's disease, or cardiac diseases, may be present. The immediate symptoms are the same as those arising from congestion. A positive diagnosis, however, cannot be made without the physical signs. There is dulness posteriorly over the lower lobes of both lungs, which was not preceded by inflammatory symptoms. The respiratory murmur is diminished in intensity, and small sub-crepitant or crepitant râles of a liquid character are heard over the same locations. There is also a cough, with a frothy, limpid expectoration.

*Treatment.*—The chief indication is to diminish the quantity of serum in the lung-tissue, and this is done by abstracting serum from the blood through the skin and bowels. If the debility is not too great, small doses of elaterium or croton-oil may be given, to produce free evacuations from the intestines. Hot-air baths, hot bottles and blankets are useful in promoting perspiration. Acetate of potash may be given to act on the kidneys and increase the flow of urine. Wet cups, applied to the chest-walls posteriorly, are also beneficial.

Dyspnœa, arising from œdema glottidis and mechanical occlusion of the air-passages, is considered in other chapters.

# CHAPTER XV.

In this affection there is an exudation of serum, under-neath the mucous membrane lining the upper portion of the larynx. Above the vocal cords this membrane is loosely attached to the underlying structures, and is more liable than other parts of the organ to be the seat of serous exudation.

The greatest amount of œdema will be found in the ary-teno-epiglottidean folds, situated at the sides of the superior aperture of the larynx, and at the base of the epiglottis. The aryteno-epiglottidean folds are reduplications of mucous membrane which loosely cover the cuneiform cartilages. Large, irregular pouches, which are here developed by the infiltration of serum, hang over the laryngeal aperture. These bags are forced in with each inspiration, making the opening still smaller, and seriously obstructing the ingress of air.

Œdema Glottidis occurs more frequently in adults than in children ; the reasons for this are—1. That in early life the mucous membrane of the larynx adheres more intimately to the adjacent tissues. An exudation of any kind from the blood-vessels would therefore appear on the free surface of the membrane, and not on its attached portion ; 2. The diseases which occasion œdema are more common in advanced life than in youth.

The affection depends on conditions which give rise to exudations of serum in other parts of the body, such as obstructions to the circulation; inflammations, lack of tonicity in the vascular walls, or a watery condition of the blood. It is not unusual during the progress of all chronic kidney-diseases, erysipelas, small-pox, continued fevers, etc. It is in most cases an attendant of acute and chronic inflammation of the larynx; it may arise, however, as an independent affection. When it proceeds from inflammation, Virchow applies to it the term collateral œdema. The inflammatory *stasis* offers an obstruction to the circulation in the diseased part, increases the pressure in the blood-vessels, so that the watery portions exude in the areolar tissue. Exceptionally, it has been known to occur in thoracic aneurism, and in quinsy sore-throat, and pharyngitis from extension of the inflammation. Whether occurring alone, or in connection with local or constitutional diseases, the symptoms of œdema glottidis are distinctly marked. The patient complains of great difficulty in breathing, which seems to proceed from an obstruction located in the throat, and he coughs violently in order to eject it. If the epiglottis be involved to any extent, there will be pain in the act of swallowing. The difficult respiration rapidly increases. Extreme distress is apparent. The patient grasps the throat violently, in vain endeavors to relieve himself, and begs and prays for help. The respiration is hard and rasping in character. The voice is usually husky, but it may be clear if no inflammation is present. More difficulty is experienced during inspiration than with expiration, owing to the fact that the pendulous bags of serum at the edge of the larynx are forced down by the current of air, and almost completely close up the canal.

The expiratory act will be found comparatively free. If laryngeal inflammation be present, both inspiration and expiration will be difficult. On examination of the throat the epiglottis may be seen enlarged and prominent, and, if the finger be carefully inserted, the puffy, œdematous swelling is readily felt. If the symptoms are not relieved, the patient soon dies asphyxiated. The duration of œdema glottidis is variable. It may destroy life in a few moments, or it may last for hours before a fatal termination.

*Treatment.*—There is no time for vacillation in these cases. Some measure for relief must be instituted without delay. Should the affection be complicated with laryngitis, and the dyspnœa not very urgent, a brisk cathartic may be given, and leeches may be applied to the top of the sternum, and at the sides of the neck. Leeches should never be applied directly to the larynx in inflammation, as a great deal of local œdema generally follows the bite.

In the majority of cases this kind of treatment will not avail much; operative measures have to be resorted to. Local scarification, as employed by Dr. Buck, of this city, is highly recommended. In performing this operation, a curved bistoury, covered almost to the point with adhesive plaster, is used. The forefinger of the left hand is passed down to the back of the tongue until the swelling is reached. The knife is then introduced, following the finger as a guide, and the bags of serum are punctured. Great care must be taken not to wound any part but the œdematous stricture, or the flowing of blood into the larnyx may choke the patient before the œdema is removed.

Scarification is sometimes rendered extremely difficult, because of the efforts at vomiting induced by the irritation of the finger in the throat. In such cases perseverance ceases to be a virtue, and tracheotomy or laryngotomy should at once be performed (*see* pages 89, 90). Either of these operations **may** be performed in all serious cases.

# CHAPTER XVI.

## CONVULSIONS.

*Synonymes.*—Eclampsia, Fits, Falling-Sickness, Spasms. A convulsion is an involuntary contraction of one or more muscles, with or without loss of consciousness. The sensorial and intellectual faculties are seldom affected except in general convulsions. The muscular contractions may be either tonic or clonic. In the former the spasm is continuous, in the latter each contraction is followed by relaxation. The spasmodic movements succeed each other with rapidity. Tonic contractions appertain especially to tetanus. The clonic variety is peculiar to epilepsy and all other classes of convulsions.

Convulsions depend either on an irritation transmitted from the periphery to the nerve-centres, or on an abnormal irritability, arising directly in the nerve-centres, which calls forth excessive and irregular action in the motor nerves.

According to Longet, sensations coming from the periphery to the brain are converted into motor impulses through the tuber annulare.

Irritation of this ganglion, whether proceeding from external sources or acting through the blood, will excite irregular muscular movements throughout the body.

Convulsions are merely symptomatic phenomena, representing diverse pathological conditions; the significance of a convulsion, therefore, depends upon its cause : it may be the premonition of death, or only the result of indigestion. Convulsions may occur at any age, but they are most frequent during infancy.

The rapidly-developing delicate tissues of the child possess a susceptibility which intensifies every irritation, and slight causes will excite irregular action and disarrange the nervous system. As children advance in years this sensibility decreases, and consequently they are less liable to convulsive attacks. In adult life, except under the form of epilepsy, they are comparatively rare.

Infantile convulsions usually occur during the first dentition and early part of that period. The first few months after birth give the greatest percentage of cases. Convulsions *in utero* have been recorded by some observers.

Children whose parents have been subject to eclamptic attacks are more liable than others to the affection. Causes insignificant in themselves develop this hereditary tendency. A debilitated state of the system is a predisposing cause. Those who have soft skulls from rachitis suffer frequently from convulsions. As exciting causes may be enumerated : indigestion, worms in the alimentary canal, teething, burns, scalds, eruptions, foreign bodies penetrating the integuments, the application of mustard-poultices, and blisters, fright, affections of the brain, such as meningitis, congestion, tumors ; exanthematous disorders ; degenerations of the kidneys, pneumonia, bronchitis, etc.

The attack in many instances can be traced to indigestion, solid food in the alimentary canal, unhealthy milk, and

12

arrow-root, or other articles partially cooked, and remaining unacted upon by the digestive fluid. An irritation is consequently produced, which is carried by the sensory nerves to the brain, and convulsions follow. Worms in the alimentary canal have a direct irritating action upon the mucous membrane of the intestines. They also diminish the digestive functions, and lower the vitality of the system; hence both causes, acting together, may excite the abnormal muscular movements.

During the first dentition, convulsions are remarkably frequent. In fact, the great majority of diseases peculiar to infancy develop during the evolution of the teeth. At this time the swollen and tender gums give rise to constant irritation. The child becomes fretful and feverish, and if there happen to be a very slight predisposition to convulsive attacks we may depend upon their occurrence. Convulsions proceeding from the reflex irritation of teething are said to be more serious than other varieties, and paralysis is not an uncommon sequence.

Irritating applications to the integument, in the form of blisters or mustard-poultices, are attended with danger. Great care should be exercised in their application. A blister scarcely two inches square may cause alarming attacks.

Diseases of the brain in children are usually marked during some part of their course by convulsions. In acute hydrocephalus they occur in the later stages of the disease—exceptionally they appear in the first stage.

Many of the narcotic medicines cause convulsions. Poisoning by stramonium-seeds is not uncommon. The only reliable test of this occurrence is the presence of the seeds in the matter vomited.

Convulsive movements may affect all the muscles of the body, involuntary as well as voluntary, or be limited to a single muscle, or to one set of muscles; one side of the body may alone be convulsed, or alternate convulsions of each side, or of different limbs, may take place.

In the affection known as *inward convulsions* the diaphragm, the muscles of the abdomen and thorax, and occasionally the muscles of the larynx, are involved.

The symptoms of eclampsia can conveniently be divided into premonitory and immediate. The premonitory signs, however, are not always present.

For a variable length of time preceding the fit, the child may be feverish and restless. The sleep is disturbed, and muscular twitchings are observed. If teething, the child moans, moves its head about, and the jaws are worked from side to side. If undigested food or worms are present, there will be a tympanitic abdomen, and eructations of gas from the intestinal canal. In brain-affections, the abdomen is flattened; there may be vomiting, projectile in character, and without nausea. There is pain in the head, and, when carried rapidly from one place to another, the child screams violently.

The convulsive movements commence suddenly. The child cries sharply, and falls. The muscles for a moment become rigid. The corners of the mouth are drawn down, the eyes are either fixed or oscillating, generally the former. There may be either convergent or divergent strabismus. Respiration ceases. The child's face, which was at first pale, becomes livid, the veins of the face and neck are turgid and filled with blood, and a gurgling noise is heard in the throat. The rigid condition of the muscles, or tonic contractions, continue but a few seconds, and they are succeeded by alternate contractions and relaxations, or clonic spasms. The limbs are moved violently about, rapidly extended and flexed. These clonic movements cease, and the patient sinks into a deep sleep or a semi-comatose condition.

The convulsive movements in children usually continue longer than in adults. The whole paroxysm lasts from half a minute to two minutes, or even longer. The fits may succeed each other with such frequency as to seem continuous, but this is rare. The immediate effects produced by the muscular contractions are worthy of notice. They may be witnessed in all kinds of convulsions. The abdominal muscles, by pressure on the intestines and bladder, may expel the fæces and urine. It is not unusual for a fit to terminate in this manner. The spasm of the respiratory muscles, including those which govern the glottis, prevents ingress and egress of air, and a partial asphyxia is the consequence. The pressure of the muscles at the base of the neck, and the non-expansion of the chest, by preventing the venous blood from leaving the head, cause congestion of the brain. The muscles which act upon the tongue protrude it from the mouth. When this occurs during the spasmodic action of the muscles of mastication, the tongue is caught between the teeth and severely lacerated. Spasm of the vessels of the pia mater is said to produce insensibility.

All the symptoms described are common to true epilepsy, and it is impossible to distinguish them during the fit. In infantile convulsions the period of spasmodic action is continued over a greater length of time than in true epilepsy. The history of the case will be of assistance in determining its true nature. For instance, in epilepsy, wo would probably learn that the patient had had fits before, coming at comparatively long intervals, and without apparent cause. In the other case there would be evidences of worms in the alimentary canal, of indigestion, or some of the other special causes previously enumerated. Again, the occurrence of attacks rapidly following each other would be rather strong evidence that they were not epileptic.

A rigid condition of one or more muscles, after consciousness is restored, is an unfavorable sign, often indicating injury to some part of the brain or spinal cord. These convulsions usually cease when the exciting cause is removed, but the possibility of a fatal termination must not be overlooked.

Convulsive attacks may occasion death—1. By asphyxia; 2. Congestion of the cerebrum, or other injury to the nerve-centres; 3. Syncope; 4. Gradual exhaustion from successive or protracted convulsions.

*Post-mortem* appearances are of little value in determining the causes of the affection. The congestion of the brain and spinal cord, which we find, is probably the result of the convulsion, and not its cause.

Among the varied sequelæ of infantile convulsions we find paralysis of different parts. It may appear in one limb, or in one set of muscles, or may involve the lower half or lateral half of the body. Recovery from it is rare. Convergent and divergent strabismus likewise occur, the latter most frequently. Idiocy may result from continuous convulsions.

A loss of coördinating power in the muscles which

produce articulate sounds sometimes occasions stammering.

Amaurosis and deafness also occur. Very little can be done to relieve them.

*Treatment.*—The preventive treatment consists in attending to the general health of the child, and placing it under proper hygienic influences. Its food should be of good quality, its nurse healthy, the sleeping-apartment well ventilated, the clothing loose and not heavy. If worms are present, they must be removed by anthelmintics. Indigestion should be relieved immediately by the ordinary means. Sores or ulcers, of the integument are treated with emollient applications, and with sedatives internally.

During the paroxysm, efforts are made to relieve the severity, and as far as possible prevent a recurrence of the attack.

The child should at once be stripped and immersed in a hot bath. A tablespoonful of mustard added to the water will increase its efficacy. The child may remain in the bath from two to four minutes at a time. Some recommend firm pressure around one arm and leg on opposite sides of the body. This procedure is of benefit in that variety of spasm called by Trousseau *tetany;* but in this affection it would be of little service. As soon as the paroxysm has ceased the bowels should be emptied with castor-oil, or by injections of warm water. After the evacuation the following may be administered, by enema—

℞. Misturæ assafœtidæ  .  .  .  .  .  fl. ʒ ss.
Aquæ .  .  .  .  .  .  .  .  fl. ℥j. M.

and repeated when necessary. Bromide of potassium, in one

or two grain doses, is also a valuable remedy. The dose of this, may be increased if desired. Should the convulsions be violent, protracted inhalations of chloroform may be employed, and repeated with benefit.

Convulsions arising from cerebral lesions, such as inflammation, etc., will not give way to the treatment recommended. This variety might as well be let alone, as it usually terminates fatally.

In all convulsive attacks a rigid investigation into the cause of the convulsion should be instituted, and treatment directed to its removal should be commenced without delay.

### CONVULSIONS IN THE ADULT.

Convulsions in the adult acquire an importance which they do not possess during infantile life. In many cases they indicate the presence of constitutional lesions, which may bring about a fatal termination in a short period. An extended description of the diseases which give rise to these convulsions is, with the limited space at command, inadmissible. All the prominent features of each condition, and especially the different signs which lead to a correct diagnosis, will, however, be fully considered. These points of difference cannot be too closely observed, and they should be studied more carefully than the points of resemblance.

These convulsions may be classed under five separate heads: 1. Those which arise from the retention of urea in the blood in disease of the kidneys, viz., uræmic convulsions; 2. Convulsions which characterize epilepsy; 3. Those rising from affections of the brain, such as extravasations of blood in its substance, or upon its surface; 4. Hysterical convulsions, and 5. Convulsions due to the excessive use of alcohol.

1. URÆMIC CONVULSIONS.—In the chapter on uræmic coma, the source and character of the poison (*urea*) which accumulates in the blood in Bright's disease of the kidneys were fully considered. It is said to act on the base of the brain and medulla like any other irritant, calling forth irregular and violent muscular movements.

These convulsions may also be due to œdema of the brain-substance, which exists in common with œdema of other parts in Bright's disease (*Roberts*). The pressure of the effused serum empties the arteries, and diminishes the amount of blood in the organ.

Preceding the commencement of the convulsion, the patient complains of headache, dimness of vision, dizziness and other symptoms referable to the nervous system. The stomach is irritable, and the bowels are usually relaxed. The countenance has a pale, waxy appearance. There is œdema under the eyes. Pressure on the lower limbs may leave a pit or indentation under the finger, showing the presence of œdema. Coma may or may not occur before the paroxysm. The urine may be scanty, and of a high color.

It must not, however, be forgotten that uræmic convulsions, occurring with the small contracted kidney, may have none of these characteristic symptoms of diseased kidney preceding them.

The paroxysm appears suddenly. The body and extremities become violently convulsed. Spasmodic contractions of the clonic variety succeed each other rapidly. The face becomes livid, the eyes are glassy and fixed, or may oscillate

CONVULSIONS.                          185

from side to side (*nystagmus*). The pupils are contracted
or dilated, usually the latter. Froth, mixed sometimes with
blood, collects around the mouth, and in exceptional cases
the tongue may be bitten. There is a strong urinous odor
emanating from the perspiration. When the convulsions
cease, the patient sinks into a deep coma, which usually
ends in death. There may be only one convulsion, or the
convulsions may succeed each other at short intervals for
several hours. The points of difference which distinguish a
uræmic convulsion from epilepsy, or from apoplectic convulsions, require careful investigation.

In uræmic convulsions both sides of the body are equally
affected by the spasmodic movements. In epilepsy one side
is convulsed more violently than the other. There are few
exceptions to this rule. In uræmia we find œdema of the
face and extremities, and urinous odor to the perspiration,
which are generally absent in cerebral extravasation and in
epilepsy. A chemical and microscopical examination of the
urine will probably show, in uræmia, albumen, and fatty,
granular, or hyaline casts, while in epilepsy and cerebral
extravasation they are usually absent. In one case we have
an antecedent history of Bright's disease of the kidneys;
in epilepsy a history of previous convulsions, with perfect
health during the intervals. The tongue is generally bitten
in true epilepsy, rarely in a uræmic convulsion. Following
the latter, there is deep coma; in the former merely a deep
sleep, from which the patient may be aroused. In cerebral
extravasation there is paralysis with irregularity of the
pupils, which is not present in uræmia. In the former also
there is sometimes rigidity of the muscles following the
attack; in the latter, this is rarely manifested. The treat-

ment of uræmic convulsions is similar to that pursued in
uræmic coma (*see* Coma).

PUERPERAL CONVULSIONS.—Convulsive attacks are not
unusual during the period of utero-gestation, particularly
toward its termination. They may arise from hysteria,
epilepsy, etc., but the vast majority are due to uræmic
poisoning. The enlarged uterus presses upon the blood-
vessels of the kidney, causing congestion of that organ, and
subsequent retention of *urea* in the blood. For a variable period previous to the convulsive seizure, the woman
may present all the ordinary signs of Bright's disease (*see*
Uræmic Coma). The convulsion is similar in all its features
to that previously described.

The seizure may cause the death of the child *in utero*.
The placenta may be compressed, so as to prevent the fœtal
blood from being aërated, or the child may be poisoned by
the urea, and die in a convulsion.

*Treatment.*—Inhalations of chloroform are employed to
stop the convulsion. Should the attacks continue, premature labor must be induced, and the uterus emptied of its
contents. If the cervix is undilated, sponge tents may be
inserted. When these have enlarged the canal somewhat,
Barnes's dilators are passed up, and distended with water
to such an extent as to thoroughly dilate the cervix. A
catheter introduced between the membranes and walls
of the uterus is sometimes employed to hasten delivery.
When the cervix has been sufficiently dilated, the child is
delivered by version, or with forceps.

The subsequent treatment consists in eliminating the
poison from the blood of the patient, and building up the
health by tonics and good diet.

EPILEPTIC CONVULSIONS are more common than any other variety. They may arise at any period of life. The largest proportion of cases, however, occur between the ages of ten and twenty (*Reynolds*). But little is known as to the pathology of the disease. Among the numerous causes given are: 1. Cerebral anæmia arising from spasmodic contraction of the vessels which supply the brain, diminishing the quantity of blood going to that organ. 2. Irregular distribution of blood to the brain, giving an over-supply to one part of the organ, and too little to another, exalting the excitability in one portion, and diminishing it in the other. 3. Excessive sensibility and excitability of the medulla oblongata, with or without spasm of its vessels (*Hammond*). 4. Softening of the pituitary body. 5. Induration of brain-substance; and, 6. Thinning and dilatation of the cerebral blood-vessels, with resulting anæmia, and exalted excitability of the medulla.

Epilepsy is often connected with masturbation, venereal excesses, syphilis, cerebral tumors, fright, etc., etc.

How far venereal excesses and syphilis tend to develop the disease is uncertain, unless by increasing the general excitability of the nervous system, and by lowering the general health.

Cerebral tumors excite convulsions by direct irritation, but we cannot place them under the head of true epilepsy any more than those arising from cerebral extravasation, or uræmia.

Many authorities give two varieties of true epilepsy: a mild form (*le petit mal*), where there is sudden unconsciousness, and little or no spasm; and *le haut mal*, where the loss of consciousness is complete, and the convulsive move-

ments general.  It is very evident that there are two forms
of epilepsy, differing in severity, but we can hardly apply
the term epilepsy to every slight loss of consciousness, or
"absence," without convulsive movement.  Many persons
have moments of partial unconsciousness, who have never
had muscular twitchings of any sort, and who are free from
hereditary taint.  These persons are anæmic, dyspeptic, or
both, and the attacks partake more of the nature of syncope
than any thing else.  I am acquainted with a gentleman
who is affected suddenly once or twice in the month with
partial or complete unconsciousness.  It always takes place
immediately after a hearty dinner, and is without spasm of
any kind.  Occasionally it is connected with a little vertigo.  Such cases should not be classed under the head of
epilepsy.

A true epileptic attack is commonly preceded by a warning called the epileptic aura.  Strictly speaking, this term
does not apply to all varieties of altered sensation which
give notice of the coming fit, but only to those which give
the feeling of a wind or breeze blowing on the person.
However, as it is in common use, it will be retained in this
connection.  This premonitory symptom assumes different
forms.  Sometimes it consists in a general feeling of weakness, or of unpleasant sensations in the epigastrium or
head.  It may be a sharp pain in one extremity or the
other, which seems to extend upward until it reaches the
head, when the paroxysm appears.  These warnings are not
present in all cases.  At the commencement of the attack
the patient usually utters a loud cry, and falls suddenly to
the ground, completely unconscious.  The countenance is
pallid.  All the muscles are fixed in a tonic spasm.  The  ·

pulse sometimes cannot be distinguished at the wrist, owing to the contraction of the muscles. Respiratory movements have ceased. The eyes are fixed, the pupils dilated. Some say that the pupils are contracted in the early part of the stage, but this is doubtful. This condition of tonic spasm lasts from ten seconds to half a minute, when the clonic spasms commence. The countenance is now engorged with blood and livid. The blood-vessels of the face and neck are distended enormously. Bloody foam collects around the mouth. The eyes roll from side to side. The pulse is full and labored. The clonic stage continues from thirty seconds to one minute. All the muscles then relax and the patient sinks into a deep sleep, which may last several hours. In these typical cases of epilepsy the patient is entirely without knowledge of the fit when consciousness is restored. Sometimes epileptic fits take place during the night and continue for some time, the person being utterly ignorant of them. He only knows that he wakens in the morning with sore limbs and wounded tongue. These night-fits are apt to be milder in form than those occurring during the waking hours.

The sequelæ of epilepsy are idiocy and insanity. Long-continued attacks are often followed by either one or the other of these affections. When they reach this point, very little can be done to remove the disease. A fatal termination is so extremely rare in epilepsy that we are not in possession of any peculiar or characteristic *post-mortem* changes. The points of difference between an epileptic convulsion and one arising from uræmic poisoning have already been given. Epilepsy is easily diagnosed from hysteria. In epilepsy there is complete unconsciousness, and the patient falls, wherever she may be, sometimes into the fire or down the stairs. In hysteria the patient knows every thing that is going on, as can be ascertained by watching the eyes; and she will fall in a soft, comfortable place, where there is little danger of receiving injury. Hysterical spasms are not so violent, nor is the tongue bitten, as in epilepsy. The face is not livid, and usually there is a choking sensation as if a ball were rising in the throat.

These convulsions are sometimes feigned by a class of persons called *malingerers*. Such cases are recognized by the fact that respiration does not cease, nor is the tongue bitten. The malingerer never falls where he is likely to hurt himself, and threats to use hot irons or hot water will bring about a speedy recovery.

From apoplexy it is distinguished by the absence of irregularity of the pupil, of paralysis, and also by the fact that the subsequent coma is complete.

CEREBRAL EXTRAVASATION.—Convulsions from this cause are extremely rare.

The patient previous to the convulsion may be affected with muscular twitchings about the face or slight numbness in one of the extremities. He may complain of a "fulness" about the head, and severe pain. The fit comes on suddenly, at the time of the extravasation. Convulsions from cerebral extravasation resemble the convulsions already described, in all the main features and symptoms.

The pupils are usually irregular, one contracted and the other dilated, or they may be both dilated.* There is always paralysis, generally of one lateral half of the body;

* There is an exception to this in extravasation of blood into the pons Varolii. In that case, the pupils are markedly contracted.

but this is not clearly manifested until the subsidence of the convulsion. When the spasms have ceased, the patient exhibits all the signs of compression of the brain—such as deep coma, slow, full pulse, dilated pupils—and he cannot usually be roused from his stupor. In epilepsy the patient is easily aroused.

The absence of albumen and casts in the urine, and of œdema of the extremities, will be sufficient in most cases to exclude uræmic poisoning. The fact, however, of the occurrence of Bright's disease in connection with apoplectic extravasation must not be overlooked. Such cases are not unfrequent. The presence of paralysis will under such circumstances lead the practitioner to the real seat of the lesion.

*Treatment.*—If the patient is full-blooded and plethoric, and the pulse full and hard, the abstraction of nine or ten ounces of blood from the arm will be decidedly beneficial. Even if it does not relieve in a marked degree the severity of the convulsive attacks, it will lessen the intra-cranial congestion, and thereby the danger of further extravasation.

When the patient is not plethoric, and when other diseased conditions tend to decrease the vital force, bloodletting should be avoided. The treatment in such cases is limited to the prevention of inflammation, absorption of the clot, and restoration of power to the paralyzed parts (*see* article on Coma).

RUM CONVULSIONS.—RUM EPILEPSY.

Persons who indulge freely in alcoholic stimulants not unfrequently suffer from spasmodic attacks resembling those of true epilepsy. The affection arises probably from irritation produced in the nerve-centres by the alcohol, and also from congestion of the same parts. Much difficulty is encountered during the attack in distinguishing its true character. It will be found, however, that the tongue is not bitten, nor is one side of the body more convulsed than the other, as in true epilepsy. The history of a long-continued " spree," and the odor of alcohol, will also serve to distinguish them.

It is also necessary to decide between these convulsions and those due to cerebral extravasation. Here, again, the presence of paralysis is an important feature. It is never found in simple rum convulsions. Following the latter there is also a stupor from which the patient is readily aroused, while in apoplexy the coma is persistent. Here the history of the case is likewise of advantage.

*Treatment.*—During the attack little is to be accomplished by treatment. Subsequently cold water may be poured on the face, and opium or bromide of potassium may be given to moderate the nervous irritability, and promote sleep.

HYSTERICAL CONVULSIONS are peculiar to young unmarried females ; but they may occur in the married state or in advanced life. Delicate women of nervous temperaments and excitable dispositions are generally the subjects. The disease is often connected with functional or organic disease of the generative organs; unsatisfied and uncontrollable passions, masturbation, etc., are not unfrequent causes.

The patient for some time previous to the attack may complain of a sensation in the throat, as if a ball were rising up and choking her (*globus hystericus*), or she may be affected with violent fits of laughter and crying, or with some of the other varied forms of hysterical manifestations. As the attack appears the patient sinks down in a comfortable spot where there is no danger of injury. The limbs are jerked about irregularly, and with less force than in an epileptic convulsion. The breathing is jerking and spasmodic; sometimes she appears as if choking. She shrieks loudly at one moment, and at another mutters incoherently; close inspection will show that the patient is not unconscious, and that the pupils are in a normal condition. There is none of that lividity of the face or distention of the blood-vessels which is characteristic of epilepsy. The paroxysm may terminate in another fit of crying or laughing, or it may be followed by sleep. Often its close is accompanied by the discharge of a large quantity of pale urine.

*Treatment.*—A pitcher of cold water should be poured slowly on the face and head. This procedure may be repeated until the convulsion ceases. Should the attack be repeated, a shower-bath will be found an excellent remedy. In very delicate females, however, this would not answer, but the cold douche to the head can be employed without injury.

The subsequent treatment has reference to the general weakened nervous system of the patient. Cold bathing, tonics, antispasmodics, good diet, and the practice of self-control, should be recommended.

TETANIC CONVULSIONS occur in tetanus. The disease arises generally from traumatic causes, such as wounds from rusty nails, etc., involving branches of nerves. Some cases arise from cold. The convulsions are caused by irritation of the spinal cord, which has been excited by injury of the peripheral nerve. They are tonic in character,

13

and extremely violent. When the muscles of mastication
are affected, the jaw is tightly closed, giving rise to *trismus*
or *lockjaw*. When the muscles of the back are involved,
the body is arched and rests on the head and heels (*opisthotonos*). Contractions of the muscles on the anterior surface bend the body forward (*emprosthotonos*), contractions
of one side give a lateral inclination, called *pleurosthotonos*.
When tetanus is once fully established, a breeze, the creak-
ing of a door, and other slight causes, suffice to excite a convulsion. Tonic spasm of the respiratory muscles generally
kills, the patient dying from asphyxia.

*Treatment.*—Anæsthetics, opiates, chloral, or assafœtida,
can be administered in large quantities.

# CHAPTER XVII.

## SUSPENDED FŒTAL ANIMATION.

Pressure on Umbilical Cord.—Injury to Brain.—Rupture of Umbilical Cord.—
Asphyxia.—Syncope.—Congestion of Brain.

DURING the progress of labor the child is subject to many accidents which may supend for a time the functions of life or completely destroy it. Thus, the umbilical cord may be pressed upon by the head in its passage through the straits of the pelvis; the cord may be wound around the neck; the air-passages filled with mucus so that the child's blood remains unaërated, and a condition of asphyxia induced.

Profuse hæmorrhage, due to rupture of the cord or to separation of the placenta, occasions another variety of suspended fœtal animation known as syncope. The head may be compressed in the maternal passages, or by instruments, with such severity as to cause congestion of the brain.

Of these three conditions asphyxia is most commonly met with. The child in this, as in the former cases, is born apparently lifeless. The face is swollen and of a dark-blue color, and the lips are livid and everted. The extremities and general surface may present a similar appearance.

Respiratory movements are absent, or there may be a slight gasp, repeated at long intervals. The pulsations of the heart are extremely feeble; as long as any movement can be distinguished, there is hope of resuscitation. A favorable result is scarcely ever obtained when the heart has entirely ceased its action. In cases where the asphyxia is produced suddenly, lividity may to a certain extent be absent, but this is rare.

In the second variety, or the state of syncope, the child is pale and cold. The lips are colorless. Respiratory movements are sighing in character or absent. The extremities are limber and flaccid. The pulse cannot be detected at the wrist, but weak pulsatory movements of the heart may be heard with a stethoscope.

When congestion of the brain exists there is some lividity about the head and face, but the color is not so dark as in asphyxia, and the capillaries of the extremities do not present the same blueness.

*Treatment.*—In all cases exertions to restore life should be made so long as the faintest movement of the heart can be detected. Life has been restored after an hour's labor, and it is not uncommon for a child to remain for half an hour without breathing, and yet be finally restored. Even when respiration has been established the treatment should be continued until the child cries vigorously.

In the first variety, where asphyxia exists, the child may be plunged alternately into warm and cold water to excite respiration through the sensory nerves of the cutaneous surface. Slapping the body at the same time with the flat of the hand is also beneficial. In mild cases this method alone will answer. Should they fail, artificial respiration by Sylvester's method (see chapter on Asphyxia), or inflating the lungs by insufflation, must be tried. In doing this the

mouth and throat of the patient must be cleared of mucus, the larynx pressed against the spinal column to prevent air from entering the œsophagus, while the physician,
with his lips applied to those of the child, blows steadily
into the lungs until they are expanded; when this is done
pressure is made on the lateral walls of the thorax to force
the air out. Again they are inflated and again compressed
until the respiratory movements are naturally performed.
Sylvester's method is preferred above all others.

The chief requirement in the condition of syncope is to
furnish more blood to the child. This is accomplished by
"stripping" the cord from the placenta toward the child's
abdomen, i. e., pressing the blood along the vessels to the
child. Friction and warmth to the surface are also necessary.

In the congestive variety the umbilical cord is cut at
once and allowed to bleed freely for a few minutes, while
the surface is rubbed and respiratory movements assisted by
alternate pressure and relaxation on the thoracic walls.

# CHAPTER XVIII.

Rupture of the Uterus.—Prolapse of the Funis.—Short Cord.—Irregular Presentations.—Application of the Tampon.

RUPTURE OF THE UTERUS.—Among the serious accidents to which parturient women are exposed there is not one more serious than rupture of the uterus. It is one of the worst complications of labor. The prognosis in all cases is bad. This accident is of more frequent occurrence in multipara, or those who have passed through several labors. Women in labor with the first child are not liable to it. The successive enlargements of the uterus diminish the strength and firmness of its walls, and develop a tendency to rupture.

Rupture of the uterus may occur at any period of utero-gestation, but usually it takes place during the second stage of labor. At this period the resistance to the uterine contraction reaches its maximum. The head of the child engages against the bony walls of the pelvis with considerable force. If, now, the linea ilio-pectinea be abnormally prominent and labor delayed, the contractions force the neck of the uterus against this part, and laceration results. In nine cases out of ten the rupture starts at the neck, but it may commence in other portions of the uterine walls.

Abnormal thinness of the uterine walls, and fatty degeneration of the uterine fibres, are liable to cause rupture, if there is the slightest over-distention or obstruction to the free passage of the head. Great distention from multiple fœti or monsters, even where the uterine walls are of normal thickness and structure, is an exciting cause.

Deformities of the pelvis, by obstructing the passage of the child, and increasing the internal pressure on the walls of the uterus, introduction of the hand or instruments into the uterus, are not uncommon causes. Rupture of the uterus may also arise from blows on the abdomen, or from violent straining efforts.

The dangers from rupture of the uterus are shock or collapse, hæmorrhage, peritonitis, or metro-peritonitis, and strangulation of intestines.

The principal and immediate danger arises from hæmorrhage. The flow of blood from dilated vessels of the uterus may put an end to life in a few moments. If the contractions of the uterus continue after the accident, there will be less danger of bleeding. In connection with the effects of loss of blood on the system, there is more or less danger from shock. In all injuries to internal organs this peculiar sudden loss of vitality is present. Sometimes the loss of blood is slight, but the shock is so great that the patient never rallies.

When immediate danger from hæmorrhage and shock has passed, peritonitis or metritis is apt to supervene. If the inflammation of the peritonæum be of any great extent, if it involve more than that portion covering the uterus, a fatal termination may be expected.

After the rupture has occurred, a portion of intestine

may pass through the opening, and be tightly strangulated by the contracting uterine walls. If this complication have not been recognized by the hand in the uterus, it will soon manifest itself by violent vomiting, at first of the contents of the stomach, and then of fecal matter, and by obstinate constipation, pain and tenderness over the abdomen, and finally collapse.

At the time of rupture the woman shrieks loudly, and complains of an agonizing pain in the hypogastric region. If the physician be near the bedside, a distinct " *tear* " may be heard. There is a gush of blood from the vagina, and the presenting portion of the child immediately recedes. In many cases the child can be felt in the abdominal cavity outside of the contracting uterus. The patient's countenance becomes excessively anxious and pallid. The pulse is rapid and very feeble. In severe cases the patient may succumb at once. If the patient survives the combined effects of shock and hæmorrhage, there is still very little chance of escaping metro-peritonitis or other complications of the accident.

*Treatment.*—In every case the child should be delivered at once. If the head is within reach, the forceps can be used, or version performed to effect that object. When the child has passed completely out of the uterine cavity, Prof. T. G. Thomas, of this city, recommends the performance of gastrotomy, and abstracting the child through the opening in the abdomen. He believes that the danger to the mother's life from the operation is not so great as when the child is taken out through the natural passage, because in this latter case some portions of the intestine are almost certain to be caught in the opening and strangulated; and also that an opening in the abdomen, besides obviating this danger, gives an opportunity to clean the cavity of all blood or portions of placenta which would excite peritonitis. Other authorities recommend the introduction of the hand in all cases without exception, and the delivery of the child through the natural opening. In so doing, great care should be taken to prevent portions of the intestine from being dragged through the hole in the uterus.

Stimulants are to be freely administered to counteract the effects of the collapse; styptics, to prevent hæmorrhage, and opiates in quantities sufficient to relieve pain, are always necessary.

PROLAPSE OF THE FUNIS. — When the umbilical cord enters the vagina in advance of the child's body, it is said to be prolapsed. If labor proceeds under such circumstances, the cord is compressed against the walls of the pelvis, and the aërated blood coming from the placenta cannot reach the child. If this pressure is maintained for many minutes, the child dies asphyxiated.

Prolapse of the funis occurs once in every two hundred and fifty cases (*Thomas*). It is caused by unusual length of the cord, sudden escape of liquor amnii, excessive quantity of liquor amnii, transverse presentations, and obliquity of the uterus.

If the membranes have ruptured, the cord can be recognized by its isolation from surrounding structures, and the rapidity of its pulsations. The pulsations are synchronous with the movements of the fœtal heart.

*Treatment.*—If a diagnosis is made before the head is engaged in the superior strait, the patient should be placed on her chest and knees; the hand of the attendant should then be inserted into the vagina, and the cord grasped and gradually returned to the uterus at the point where it made its exit. These efforts should be made while the uterine fibres are relaxed. The cord is retained inside the cervix by the finger of the physician until the uterus is firmly contracted. The woman should remain on her chest and knees until the head of the child is engaged in the superior strait.

This method of replacing a prolapsed cord has superseded all others. It was first introduced by Prof. T. G. Thomas, of this city.

If the child's head passes the superior strait before the prolapsus has been discovered, the forceps must be applied, and the labor completed without delay.

SHORT CORD.—The length of the umbilical cord is subject to considerable variation. Schneider reports a case in which the cord measured over three yards, and Cazeaux speaks of one which was only nine inches in length. It usually measures from eighteen to twenty-four inches.

A short cord retards the progress of labor. It may also give rise to hæmorrhage by causing premature separation of the placenta, or rupture of the cord. When the cord is shortened by winding around the child's body, similar consequences may ensue.

A short cord cannot be recognized until the commencement of labor. At this time the fundus of the uterus will be found depressed or " dimpled " with each contraction. The cervix is soft and dilated, but there is no advance in the labor. If the index-finger is applied to the child's head it will be found to recede during the relaxation of the uterine fibres. Hæmorrhage more or less profuse may also be present. (*See* Placenta Prævia.)

*Treatment.*—If the labor has not progressed beyond the first stage, the membranes should be ruptured, so as to bring the uterus in more immediate contact with the body of the child, and thus increase its power of expulsion (*Cazeaux*).

When the child's head has passed beyond the cervix, and is prevented from advancing farther by the short cord, the delivery must be terminated with the forceps. Some obstetricians advise the performance of version as soon as the cervix is dilatable.

IRREGULAR PRESENTATIONS AND POSITIONS.—In ordinary cases of vertex presentations the occiput rotates anteriorly under the pubes. Exceptionally, it rotates in a contrary direction into the hollow of the sacrum. In this position the head can only be delivered by extreme flexion. In some instances the efforts of Nature are sufficient to terminate the labor; the majority of cases, however, require the aid of the forceps.

When the patient is fully anæsthetized and in position, the male blade of the forceps, which is usually held in the left hand of the operator, is introduced on the left side of the vagina, and applied to the right of the child's head. The female blade is introduced on the right vaginal wall, and passed up to the left side of the head. When the forceps are locked, the handles should be raised toward the pubes, in order to produce greater flexion of the head. At the same time traction is made, and the head brought down to the vulva. When the head reaches this point, some obstetricians prefer to remove the forceps, and let the labor proceed naturally.

PRESENTATIONS OF THE ARM OR LEG, together with the

head, may effectually impede the progress of labor. As soon as discovered, efforts should be made to return the protruding limbs to the cavity of the uterus. Sometimes the presenting parts are so firmly wedged in the pelvic cavity, that they cannot be replaced; in such cases embryotomy or craniotomy must be performed.

In TRANSVERSE PRESENTATIONS it is not unusual for the arm and shoulder to present at the superior strait. The arm should be replaced and the head brought down (*cephalic version*). If the head cannot be brought to the superior strait, one of the lower limbs may be seized and the child delivered by *podalic version*. In the performance of version the following rules must be observed : 1. Oil the *back* of the hand and fingers only ; 2. Introduce the hand during the relaxation of the uterine fibres ; 3. Introduce the hand, which when in the cavity of the uterus will have its palmar surface in relation with the anterior portion of the child's body ; 4. Do not rupture the membranes until the hand has reached the part of the child to be brought down ; 5. The necessary manipulations *in* the uterine cavity should be made between the pains.

FACE PRESENTATIONS occur once in two hundred and fifty labors (*Thomas*). The most frequent position is the " right mento-iliac transverse." In natural labors the chin is carried forward under the pubes and is finally delivered by a process of flexion. Should the chin rotate posteriorly into the hollow of the sacrum, the longest diameter of the child's head (occipito-mental) is brought in relation with the antero-posterior diameter of the pelvis. The former measures five inches and a quarter, the latter four inches and a quarter. It is impossible, therefore, for the labor to terminate naturally. Operative procedures are always necessary.

*Treatment.*—If a diagnosis is made before the head is engaged, the face-presentation may be converted into one of the vertex by flexing the head. If this cannot be done, an attempt should be made to change the position of the face and rotate the chin under the pubes. Either the hand of the physician or the vectis may be employed for this purpose. When the movement of rotation cannot be accomplished, the perinæum may be incised and the child delivered by means of forceps. This method is recommended by Dr. Taylor. Other authorities advise craniotomy when milder measures fail.

APPLICATION OF THE TAMPON.—The tampon is employed for the suppression of hæmorrhage occurring in abortion, placenta prævia, ulceration and laceration of the vaginal walls, etc. It should not be resorted to in *post-partum* hæmorrhage.

The tampon may be made of sponge, picked lint, cotton, India-rubber bags filled with water or ice, or a surgical roller-bandage. The latter was first employed in tamponing by Prof. I. E. Taylor. He claims that the bandage is more readily introduced and removed than any other material.

Any of the substances employed may be wet in astringent solutions previous to their introduction. The operation is performed with or without a speculum. The patient should be placed in the recumbent posture and the thighs flexed on the abdomen and abducted. A speculum is then introduced into the vagina, and the lint or other materials passed up and packed tightly around and upon the cervix, increasing the quantity until the vagina is completely filled. A T-bandage is afterward employed to maintain the tampon in position. The tampon should be changed at the end of twenty-four or thirty-six hours.

When the patient desires to micturate, a portion of the plug at the entrance of the vagina must be removed. At this point the plug presses on the urethral canal, and its removal is necessary before the urine can pass through.

# CHAPTER XIX.

RETENTION OF URINE.—Retention of urine may arise from spasmodic contraction of the muscular fibres of the neck of the bladder, organic stricture of the urethra, enlarged prostate, stone in the bladder, paralysis of the bladder, abscesses in the perinæum, fracture of the pubic bones, with laceration of the urethra, and injuries to the spinal cord.

Retention which is produced by spasm of the muscular fibres accompanies exposure to cold, or acute inflammation of the urethra. It occurs suddenly, and is not connected with chronic disease of the genitals. There is pain in the perinæum and hypogastric region. If the bladder is distended with urine, a large area of dulness will be found on percussing along the pubes. Febrile excitement is also present if the retention follows inflammation.

The patient is readily relieved by the application of hot fomentations over the hypogastrium and genitals, hot baths, and by the internal administration of opium. Leeches to the perinæum are useful in some cases.

In retention from organic stricture the patient will have

had, for a variable period previous to the attack, great difficulty in micturition, a small, twisted stream of urine, and
some degree of pain. An exploration with sounds or bougies
will show an obstruction at some point between the meatus
and membranous portion of the urethra.

If the stricture cannot be dilated rapidly, and if the
condition of the patient will not permit of urethrotomy, the
distended bladder may be temporarily relieved by puncturing through the rectum. At the base of the bladder there
is a space uncovered by peritonæum, which is bounded on
each side by the vesiculæ seminalis, behind by the recto-
vesical fold of the peritonæum and in front by the prostate
gland. The operation at this point is performed by inserting the left index-finger into the rectum and carrying it
half an inch or an inch beyond the prostate, and then introducing a large, curved trochar (using the finger as a
guide) and plunging it into the bladder at that point. The
stylet is then removed, and the urine escapes through the
canula. If fluctuation cannot be detected by the finger, the
operation should not be performed.

Retention from enlarged prostate occurs in advanced life.
The hypertrophied gland may be felt by a rectal examination. If the ordinary large curved prostatic catheter cannot be passed over the obstruction, an instrument with a
shorter curve may be forced *through* the enlarged lobe into
the bladder, or the bladder may be opened through the rectum in the manner previously described.

Habitual distention of the bladder may induce a semi-
paralytic condition of the walls of the organ and produce
retention. This condition occurs not unfrequently in females
whose opportunities for emptying the bladder are often re·

stricted. It is relieved by frequent introduction of the catheter, cold hip-baths, and tonics.

When retention arises from injuries to the spinal cord the bladder should be emptied twice each day by means of a catheter, and thoroughly washed after the urine is evacuated.

DISLOCATION OF THE NECK.—This accident is usually fatal. In death from hanging the transverse ligament is ruptured, the axis is dislocated from the atlas, and the odontoid process of the former bone presses upon the upper portion of the cord. Death in such a case is almost instantaneous.

Partial dislocations of the cervical vertebra lower down are sometimes recovered from. In these cases, the head is turned to one side, and there may be slight paralyses below the point of injury.

*Treatment.*—The surgeon grasps the head of the patient, while an assistant steadies the shoulders. Extension is then carefully made, while the head is rotated toward its normal situation. Perfect rest for a few days is afterward necessary.

INJURIES FROM LIGHTNING.—The effects of lightning on the system vary in character. In some instances death is instantaneous, in others there is more or less extensive charring of the tissues, paralysis of the extremities, loss of sight, speech, and hearing, and hæmorrhage from the mucous canals. Burns produced by lightning are apt to run a protracted course, and are accompanied by extensive suppuration. Paralysis is rarely recovered from. Boudin speaks of cases where persons injured by lightning had images of surrounding objects depicted on the body and

14

clothes.    Similar curious occurrences have been recorded
by other observers.

The symptoms presented by a patient suffering from a
lightning-stroke are coldness of the extremities, sighing respiration, absence of radial pulse, and insensibility.

After death the ordinary *rigor mortis* is not witnessed,
and the blood is said to be more fluid than in death from
other causes.

The treatment consists in friction to the surface, artificial respiration, and the administration of stimulants.

Colic.—Spasmodic contraction of the muscular walls of
the intestines is generally attended with great pain.   It is
occasioned by cold, or over-indulgence in indigestible food.
It is characterized by paroxysms of intense pain over the
abdomen; vomiting is sometimes associated with it.   The
pain is distinguished from that accompanying inflammation by the fact that it is relieved on pressure.

An injection of one or two quarts of very warm water
and an opiate will cure it.   The following prescription
answers in many cases:

R. Bismuthi subnitratis    .    .    .    .    .    .    3 j.
  Morphiæ sulphatis    .    .    .    .    .    .    gr. j.  M.
  Ft. pulv. x.

One powder should be given every hour until the patient
is relieved.   Mustard or hot flax-seed poultices may also be
applied over the abdomen.   (*See* Lead Colic.)

# CHAPTER XX.

### NARCOTIC POISONS.

Opium, Belladonna, Hyoscyamus, Aconite, Tobacco, Stramonium, Chloroform, Hemlock, Lobelia, Woorara, Ether, Alcohol, etc.

OPIUM is obtained from the unripe capsules of the *Papaver somniferum*, or poppy. The juice of the capsules is the portion used. The plant is cultivated in India, Persia, Europe, and in this country. It has been employed as a medicine from the time of Hippocrates to the present day, and stands unrivalled as a remedy for the alleviation of pain.

In Turkey and China the drug is habitually smoked and chewed. In the western parts of Europe and in this country the habit of smoking and eating opium is not uncommon. It engenders exaltation of ideas, and general buoyancy of spirits. Some of the brightest lights of the literary world have fallen victims to this vile habit of opium-eating. The well-known case of Fitz-Hugh Ludlow is familiar to most American readers, and in England the celebrated Coleridge and De Quincy were victims to the drug.

The quantity of opium necessary to cause death varies with circumstances. Quantities which would destroy life in ordinary cases are eaten with perfect impunity by persons accustomed to its daily use. Enough has been taken at a dose to destroy a dozen lives. Herdouin mentions the case of a woman with cancer of the uterus who took laudanum by pints. De Quincy was in the habit of taking nine ounces daily. I have known two cases average daily from four to six ounces.

The amount which will destroy life depends also on the age of the person. Infants can bear but a very minute quantity. One drop of laudanum has been known to kill a child. Children are extremely susceptible to its influence. The smallest quantity known to have destroyed the life of an adult is two drachms of laudanum (*Skae*). In the majority of cases larger quantities are required. Opium kills in from four to twelve hours.

Some animals are scarcely affected by the drug. On apes it exerts no perceptible effect. In one instance five hundred grains were given to one of those animals without injury.

*Tests.*—Perchloride of iron gives a red precipitate with solutions of opium which contain meconic acid. Nitric acid gives a red precipitate with morphia, the principal alkaloid of opium.

The symptoms manifested in persons addicted to opium-eating are readily recognized. The face is sallow, pinched, and parchment-like. The eyes are sunken and glassy. When they are deprived of the drug there is an unsteady, trembling gait, great depression of spirits, and intense mental and physical agony. While under treatment patients endeavor by every conceivable means to obtain a dose, even getting down on their knees, begging piteously for it.

But in such cases it is rarely expedient to satisfy their cravings. " Tapering off," as they call it, will not result in cure. The appetite for the drug remains so long as they are allowed to taste and experience its intoxicating effects. Large doses of bromide of potassium will do much in these cases to diminish the craving.

The effects of poisonous doses of opium appear in from thirty minutes to two hours from its administration. Liquid preparations of opium, and the salts of morphia, act very rapidly. The patient trembles, becomes giddy, drowsy, and unable to resist the tendency to sleep. Gradually the stupor deepens, until there is perfect insensibility. The pupils are contracted, eyes and face congested; the pulse, at first rapid and small. is now slow and feeble. A marked diminution in the number of respiratory movements is discernible. From twenty per minute they run down to twelve, or even eight. The breathing is stertorous. A profuse perspiration breaks out on the surfaces. As coma deepens, and death approaches, the extremities become cold, and the sphincters relaxed. Occasionally the odor of opium may be noticed in the breath, and in such a case the diagnosis is materially assisted.

The following singular case of opium-poisoning in con-junction with cholera illustrates the characteristic effects of the drug:

A colored woman was admitted, in the summer of 1866, to the pavilion attached to Bellevue Hospital; she was suffer-ing from a bad attack of Asiatic cholera, and when brought to the ward was fast approaching a state of collapse. Inquiring into her history, she stated that the attack came on four hours previous, and while at the station-house half an hour before her admission a policeman had given her a table-spoonful of pure laudanum. As there were no symptoms to corroborate her story, I did not credit it and left her. In about three-quarters of an hour the nurse in charge informed me that the patient was insensible, and could not be roused to take her medicine. I went down immediately and found the patient as the nurse had stated, in a comatose condition. The pupils were contracted, respiration down to eight per minute. Pulse slow and small. Injections of brandy and ammonia, and strong coffee, were ordered. The body was properly stripped, and flagellation applied with twisted towels. After two hours of this treatment signs of consciousness appeared. The patient was then lifted from the bed and rapidly marched up and down the ward, supported by her nurses until she was fully restored. Five hours were spent in bringing this woman to a state of consciousness.

The treatment for opium-poisoning, and the opium itself, seemed to exert a curative effect on the cholera, and the patient was discharged three days after her admission, cured.

*Treatment.*—If the patient is seen soon after the poison has been taken, the stomach should be emptied by a stomach-pump or emetics. Twenty grains of zinc, or ipecac., a tablespoonful of mustard or common salt, will suffice to eject the poison. These medicines should be followed by copious draughts of warm water to keep up the vomiting. As soon as the stomach is emptied, belladonna, the physiological antidote for opium, may be tried. The active principle of belladonna (*atropia*) may be given by hypodermic injections. A solution of one grain to the ounce is made, and fifteen or twenty minims injected, and repeated, if necessary. Strong coffee is another antidote. In all cases the antidotes should be accompanied by stimulants. Brandy and ammonia may be frequently given by the mouth or rectum. Flagellation of the surface by the hands or towels, and causing the patient to walk about, are important aids to restoration.

The use of the Faradic current will be of service in all cases. The electrodes may be applied over the phrenic nerve and diaphragm, and over the frontal bone.

### BELLADONNA.

The leaves and root of *Atropa belladonna*, or deadly nightshade, are largely employed for medicinal purposes. All parts of the plant possess poisonous qualities. The leaves and berries are frequently eaten by children, and with deleterious effects. Thirty-six berries have produced death in a child. An infusion made from two drachms of the leaves has killed an adult. Atropia, the active principle of the plant, given in two-grain doses, has proved fatal.

The first symptoms of poisoning are dryness of the throat, constriction of the fauces, difficult deglutition, indistinct vision (*amblyopia*), or double vision (*diplopia*), headache, staggering, and confusion of ideas, stammering, etc. The pupils are widely dilated, face suffused, lips livid, and pulse rapid and intermittent. Delirium and deep coma soon supervene, followed rapidly by death. In a few cases there are convulsions.

After death putrefaction rapidly ensues. Large purple spots form on the body. There may be signs of inflammation in the stomach and intestines.

*Treatment.*—An emetic should be administered without delay, and repeated until the stomach is completely emptied. This should be followed by stimulation, friction to the extremities, and warmth. Some recommend opium as an antidote. It has been successful in one or two cases. Runge advocates the use of lime-water in large quantities as a neutralizer of the poison. Bouchard has employed the ioduretted iodide of potassium with benefit. All the strong alteratives are said to possess more or less remedial power; but experiments have not proved their efficacy.

Brandy by enema, and opium by hypodermic injection, in conjunction with large doses of lime-water, constitute the most reliable remedies that have yet been fixed upon. If the coma appear rapidly and without convulsive movements, electricity may be used with benefit, and cold water may be poured on the chest and face.

### HEMLOCK.

There are five varieties of hemlock which possess poisonous properties, viz., *Conium maculatum*, *Cicuta virosa*, *Œnanthe crocata*, *Phellandrium aquaticum*, and *Æthusa cynapium*. *Conium maculatum*, or spotted hemlock, is much used for medicinal purposes. It was a preparation of this drug which caused the death of the philosopher Socrates. All parts of the plant are poisonous. To inhale the air in the vicinity of this plant in the hot months of summer is said to be followed by slight narcotism. Its poisonous effects are manifested within half an hour after entering the stomach, and death results in from one to three hours.

The symptoms are dryness of the throat, muscular tremors, dizziness, difficult deglutition, and a feeling of great prostration and faintness. The limbs are rendered powerless, sometimes being completely paralyzed. The pupils are dilated, the pulse is rapid and small. Deep insensibility rapidly supervenes, and there may be convulsions preceding the fatal termination.

The roots of *Cicuta virosa*, or water-hemlock, are sometimes mistaken for parsnips, and eaten in large quantities. The symptoms of poisoning resemble those of the preceding variety, with the addition of vomiting, and pain in the epigastrium; convulsions are also more frequent.

The leaves and roots of the *Œnanthe crocata* are more deadly than any other species of hemlock. The plant grows at the sides of ditches and other moist places; it resembles celery.

When taken internally, it always produces violent and protracted convulsions, in conjunction with the symptoms previously enumerated (*Taylor*).

*Æthusa cynapium*, or fool's parsley, does not kill so rapidly as the other varieties. It resembles ordinary parsley, and is sometimes eaten by mistake. The symptoms commence by intense pain in the abdomen, followed by vomiting and purging, and a tendency to coma.

*Treatment.*—Empty the stomach of its contents, and use diffusible stimulants in large quantities. If there are much pain and vomiting, bromide of potassium, in ten-grain doses, may be given at short intervals.

### HYOSCYAMUS.

*Hyoscyamus niger*, or henbane, is a European plant, cultivated in this country. The leaves and seeds are largely employed in medicine. All parts of the plant are poisonous.

The seeds are more powerful than other parts. Its alkaloid (*hyoscyamia*) is a deadly poison taken in minute quantities. Animals, such as horses, goats, cows, etc., are exempt from its injurious influences, and eat it without receiving harm. Dogs and cats are soon killed by it.

Poisonous doses of the seeds or leaves are followed rapidly by dilatation of the pupils, dimness of vision, muscular twitchings, inability to articulate plainly, and a tendency to sleep. In a later stage there are vomiting and purging, abdominal pain, delirium, convulsive movements of the extremities, small, intermittent pulse, and coma, which is often followed by death.

A *post-mortem* examination shows evidences of inflammatory action in the stomach and intestines, and in a few cases congestion of the brain.

*Treatment.*—Common charcoal has been strongly recommended as an antidote by Dr. Gar. The substance rapidly absorbs the alkaloid upon which the poisonous properties of the plant depend, and prevents its peculiar action. Solutions of caustic alkalies are said to neutralize the poison. In every case stimulants should be employed, as in the other varieties of poisoning.

### ACONITE.

This drug is obtained from the leaves and root of the *Aconitum napellus* (monk's-hood, or wolf's-bane). Preparations of the leaves and root are used in medicine. The root is said to have ten times greater strength than the leaves. The plant has been mistaken for horseradish. In small doses it acts as an arterial sedative, diminishing the heart's action, and lowering the pulse. It differs from all other narcotic medicines in producing a peculiar numbness and tingling sensation in the mouth and fauces.

Cases of poisoning generally result from careless over-dosing with the tincture of the root. Thirty drops of Flemming's tincture have caused death, but there are instances of a drachm or two having been taken by mistake without fatal results. The active principle (*aconitia*) is one of the most active poisons known; one-twelfth of a grain has proved fatal.

Poisonous doses produce immediately the characteristic numbness and tingling of the mouth and fauces. The same feeling is experienced in the extremities. There are sore-throat, pain over the stomach, and vomiting. The pulse is extremely weak and compressible. The pupils are in some cases dilated, at others contracted. As in poisoning by other narcotics, there are dimness of vision, vertigo, great prostration, general loss of sensibility, delirium, and coma. Death is said to take place from syncope, asphyxia, and coma.

*Treatment.*—Emetics are first employed. Complete evacuation of the stomach is sometimes all that is required. Brandy in tablespoonful doses, given in ice-water every half-hour, is a useful method of stimulation. Preparations of nux-vomica are said to neutralize the action of aconitia. The tincture of nux-vomica has been used with apparent benefit. It may be given in ten-drop doses, every fifteen minutes, until the alarming symptoms have subsided.

### TOBACCO.

This plant was first discovered in America by the Spaniards. The English are indebted to Sir Walter Ra-

leigh for furnishing them with the "weed." The leaves
are employed medicinally as poultices to painful swellings,
and for their emetic properties. Five grains of the powder
will produce emesis. In the form of snuff it has been
employed by keepers of immoral houses to drug their victims. A teaspoonful of snuff in a glass of ale will give rise
to delirium, vomiting and purging, and faintness. The
active principle (*nicotia*) is a deadly poison. One drop will
kill a rabbit (*Taylor*). It causes death in two or three
minutes. .

The effects produced in persons of nervous temperament
by long-continued use of tobacco are well marked. An
examination of the heart shows that it is intermittent in its
action, and its pulsations more rapid than normal. The
pulse is weak. Shortness of breath and palpitation of the
heart are complained of in going up-stairs. Slight excitement induces great tremulousness. There is often impairment of the mental faculties, such as defective memory, etc.
The countenance has a sallow aspect. Some impairment
of the digestive functions is almost always present.

The effects of large quantities of tobacco on the system
are well known to smokers and chewers. Early efforts in
acquiring the habit are characterized by poisonous symptoms. There are intense nausea and vomiting. The nausea
is said to resemble that occurring in sea-sickness. Vertigo,
muscular weakness, and intense prostration verging on syncope, are also present. Later the. extremities become cold
and clammy, and convulsions sometimes precede death.

*Treatment.*—Hot bottles and blankets should be applied
to the body. Brandy by enema is always required if the
liquid cannot be retained on the stomach. Sub-nitrate of

bismuth in ten-grain doses, continued with one-fifteenth of a grain of morphia, will do much to allay the distressing nausea.

### DIGITALIS

Is a product of the *Digitalis purpurea*, or purple foxglove. It exerts a powerful sedative effect upon the heart, acts on the kidneys as a diuretic, and on the brain as a narcotic. Some ascribe its influence in diminishing the pulsations of the heart in febrile diseases to a stimulating effect on the heart's fibres, which gives them renewed vigor.

It is dangerous on account of its accumulative effect. It may be administered for several days without apparent action of any kind, when suddenly the patient is prostrated with all the symptoms characterizing poisoning by this drug. The alkaloid *digitalia,* when boiled with sulphuric acid, is changed into glucose, or grape-sugar (*Kinsman*).

The symptoms produced by poisonous doses are loss of strength, feeble and fluttering pulse, faintness, nausea and vomiting, and stupor. The body is bathed in cold perspiration, the pupils are dilated, the breathing is sighing and irregular, and convulsions are sometimes present.

*Treatment.*—Ammonia, given internally in frequently-repeated doses, is an admirable remedy, when the patient is in a state of syncope. The medicine should also be applied to the nostrils. Brandy internally, and warmth to the surface, are followed by good results.

### STRAMONIUM.

The common name of the plant is thorn-apple, or Jamestown weed. It grows all over this country, particularly along the roadsides and in moist grounds. All parts of the plant are poisonous. The seeds are not unfrequently eaten by children. These seeds are recognized by their dark, almost black color, their flat, roughened surface, and kidney-shape. The drug is much used in asthma and other spasmodic affections. Cigarettes made of the leaves are smoked by asthmatics with great relief. The active principle (*daturia*), given in small doses, proves rapidly fatal.

The symptoms of poisoning are dryness of the throat, thirst, delirium, convulsive movements, swelling of the face, dilatation of the pupil, suffusion of the eyes, small, rapid pulse, hurried breathing, and hot skin. In some cases there are pain over the stomach, and vomiting. Convulsions are nearly always present, and are liable to be mistaken for those arising from uræmia or epilepsy. On examination of the vomited matters, the seeds of stramonium will probably be discovered, which will make the diagnosis clear.

*Treatment.*—Opium, stimulants, and alkaline medicines are employed in the same manner as after poisoning by belladonna.

### LOBELIA INFLATA

Is used in medicines as an emetic and antispasmodic. The common name is Indian tobacco. It is often administered by quacks who style themselves "vegetable doctors," and is sometimes given in dangerous doses. Taylor recites several cases where death resulted from improper quantities administered by those men.

In large doses it induces excessive vomiting and purging, pain in the bowels, contraction of the pupils, delirium, coma, convulsions, and death.

The *post-mortem* appearances consist in congestion of

the membranes of the brain, and evidences of inflammation of the stomach and intestinal canal.

The treatment is confined to stimulants, and counter-irritation over the stomach.

### COCCULUS INDICUS

Contains a peculiar active principle, called *picrotoxia*, to which its poisonous character is due. The drug is sometimes given to certain kinds of fish in India to render their capture an easy matter. The seeds are small, and about the size of a pin-head. The active principle is said by Glover to produce the same class of convulsive movements witnessed after lesions of the corpora quadrigemina and cerebellum, viz., tonic spasms, and wheeling and backward movements of the body.

The symptoms and treatment are the same as in other varieties.

### MUSHROOMS.

This plant is eaten in large quantities in all parts of the civilized world. There are numerous varieties of the plant, some harmless in their nature, and others highly poisonous. Strangely enough, many which are regarded as deleterious in one part of the world are eaten with impunity in others. Mushrooms which are considered dangerous in England and in this country, are used as food in Russia; and some which are eaten in England are thought poisonous at Rome.

The poisonous mushrooms may be recognized, according to Chrystosin and M. Richaud, by their dark color, acid, bitter taste, pungent odor, and by the fact that they generally grow in damp, dark places

When a poisonous mushroom is taken internally, it causes extreme muscular weakness, vertigo, mental hallucinations, stupor, and in a few instances violent vomiting and purging. Recovery is not unfrequent, even when large quantities have been eaten.

*Treatment.*—The stomach and bowels should be acted upon by emetics and cathartics, where vomiting and purging are absent. Castor-oil, however, may be given in all cases. Opiates are recommended by some where there is much delirium without stupor. Ether has been used with benefit. If the prostration is great, the free use of diluted stimulants will be necessary.

### YEW-TREES.

The leaves and berries of this tree are extremely poisonous. An infusion of the leaves is often administered in this country to bring on the menstrual flow, or to produce abortion. Its action in this respect is not well understood. Children are often poisoned by the berries.

The symptoms are vomiting, convulsions, dilated pupils, and coma, which usually ends in death.

Stimulants are principally to be relied on in the treatment.

### CAMPHOR

Is a concrete substance obtained from the *Camphora officinalis,* an evergreen tree of China and Asia. It rarely produces death. Taylor relates the case of a man who, in twenty minutes after taking the drug, was seized with vertigo, dimness of vision, and convulsions. The pulse became rapid and weak, the extremities cold. The stomach was emptied by a stomach-pump. He suffered for a week subsequently with exhaustion, and from suppression of urine.

In some cases there are pain in the back, and rapid insensibility.

The breath of a person poisoned by camphor smells strongly of the drug, and thus the diagnosis is readily made.

*Treatment.*—Free emesis should be procured without delay. Stimulants are always necessary.

## ALCOHOL.

Large quantities of alcohol, in the shape of whiskey, brandy, etc., have produced sudden death in young persons unaccustomed to the poison. Convulsions and coma are not unfrequent accompaniments of excessive indulgence in ardent spirits (*Taylor*). (For characteristic appearances and treatment, *see* Convulsions.) Chronic poisoning by alcohol is recognized by the bloated countenance, blood-shot eyes, general tremulousness, and delirium tremens.

The treatment for this condition consists in total abstinence from liquor, and the administration of bromide of potassium.

## CHLOROFORM.

This substance is one of the most effective anæsthetics known. Its formula is $C_2HCl_3$. It is technically known as the terchloride of formyl. It is prepared by the action of chlorinated lime on wood-spirit. When inhaled, it first acts as a stimulant, causing great excitability and intoxication, then mental hallucination and delirium, and finally perfect insensibility and coma.

15

In the third stage, when the inhalations are carried beyond a certain point, the pulse becomes very small and intermittent, respiration slow, irregular, and difficult; face congested, and lips livid.

If organic disease of the heart exist, very small quantities may produce death. Sometimes respiration is suddenly suspended, and death ensues rapidly. In one instance I have seen it produce convulsions. Chloroform kills by asphyxia, syncope, or coma. After death the lungs are congested and filled with dark blood.

*Treatment.*—Artificial respiration is the main reliance in the treatment of chloroform-poisoning. Marshall Hall's or Sylvester's method will answer (*see* chapter on Drowning). Inhalation of pure oxygen is always beneficial. In some cases it may be forced into the lungs through an opening in the trachea. Slapping the patient, and pouring cold water on the surface, are also recommended. Galvanism has been successful in restoring life in one or two cases. Some rely solely on electrical stimulus in the treatment.

## ETHER ($C_4H_5O$)

Is manufactured by the action of sulphuric acid upon alcohol. The acid merely removes the water from the alcohol, to form the ether. The action of the vapor of ether is similar to that of chloroform. Its effects are, however, manifested more slowly; the resulting anæsthesia continues longer, and larger quantities of the drug are required to produce the same degree of insensibility.

The symptoms accompanying poisoning by ether are the same as are witnessed in chloroform-poisoning, and a similar treatment must be pursued.

## CHLORAL.

This drug has lately come into general use as an anodyne and hypnotic. It is made by the action of chlorine gas on alcohol. It is used in the form of a hydrate. When taken into the system it is changed into chloroform by the action of the soda of the blood.

Its poisonous influences are manifested by laborious and irregular breathing, congestion of the face, rapid and feeble pulse, numbness, and insensibility. In some cases there is considerable disturbance of the mental faculties.

After death the same lesions are found as exist in poisoning from chloroform.

*Treatment.*—Some recommend hypodermic injections of strychnia as an antidote. Artificial respiration, inhalations of oxygen, and stimulation are mainly to be relied on. Electricity is also beneficial.

## HYDROCYANIC ACID.

The common name of this drug is prussic acid. It is obtained from bitter-almonds, peach-kernels, cherry-laurel, prunus Virginiana, and bitter cassava. It is formed in bitter-almonds by the reaction of a peculiar principle called amygdaline, and water. The change is excited by the presence of a nitrogenized body called emulsine.

The essential oil of bitter-almonds is employed as a flavoring extract. Almond-water and laurel-water are used for a similar purpose.

Prussic acid is manufactured by the action of sulphuric acid upon ferrocyanide of potassium, or by the action of muriatic acid upon the cyanide of silver.

The acid obtained by this process is in a dilute form, and contains about two per cent. of the anhydrous variety. It is colorless, and possesses a peculiar odor resembling peach-kernels or almonds.

It is one of the most deadly substances known, killing more rapidly, and affording less opportunity for recovery, than any other poison. Inhalation of its vapor in a concentrated form has in some instances produced almost instant death. Scheele, while pursuing his chemical investigations with this drug, died instantly by inhaling his own preparation of it. A single drop of the anhydrous acid placed on the tongue will kill instantly. A drachm of the dilute acid will destroy life in a few seconds, unless immediate efforts at restoration are made. The poison acts as rapidly if placed in a wound. In some instances life is prolonged for three or four minutes when poisonous quantities are swallowed. In one or two rare cases a fatal termination did not occur for an hour after the administration of the poison.

*Tests.*—Taylor mentions three principal chemical tests : 1. Nitrate of silver, which gives a white precipitate of the cyanide of silver; 2. On the addition of potash, and a solution of the sulphate of iron, there is a brownish-green precipitate, which changes into blue, upon the addition of diluted muriatic acid. The blue substance thrown down is ferrocyanide of iron, or Prussian blue; 3. Bihydrosulphate of ammonia, when added to the suspected solution and warmed, makes the mixture colorless, and after evaporation leaves sulphocyanate of ammonia, which is recognized by the " blood-red " color produced by adding a solution of the colorless persulphate of iron.

When large doses of the drug are taken, the patient falls unconscious to the ground, the face becomes congested, respiratory movements labored, and diminished in length and frequency; pupils dilated, eyes glassy and prominent, pulse imperceptible, skin clammy and cold. Foam collects on the lips, the jaw drops, and death supervenes. If small quantities are taken, and the symptoms develop more slowly, there are difficult and convulsive efforts at breathing, the movements occurring at long intervals, vertigo, oppression over the precordial region, muscular weakness, and paralysis (*Bacher*). The eyes are prominent, and there are sometimes convulsive movements, and loud cries from the patient.

The *post-mortem* appearances vary. The peculiar almond ¯ odor is nearly always exhaled from the body. The lungs, brain, liver, and kidneys, are filled with dark fluid. The eyes are remarkably bright and staring. In some instances the muscles will not respond to galvanic stimulus.

The symptoms appertaining to poisoning by almond-oil, cherry-laurel, or cyanide of potassium, are developed more slowly than the preceding. Their main features and treatment are alike.

*Treatment.*—Chlorinated lime in solution, chlorine-water, or ammonia in vapor largely diluted, are good antidotes. Another method employed is to change the prussic acid in the stomach into Prussian blue. According to the " United States Dispensatory," this is done in the following manner: Ten grains of sulphate of protoxide of iron and one drachm of Tr. ferri chlor. are added to an ounce of water, and twenty grains of carbonate of potassium to one ounce of water in another vessel. The latter solution is swallowed first, and immediately followed by the preparation of iron. Cold water poured from a height upon the face, chest, and abdomen, and artificial respiration, are also recommended as efficacious remedies.

### WOORARA.

The source of this poison has been the subject of considerable controversy. Schomberg thought it was a product of a plant called *Strychnia toxifera*. Nothing analogous to the action of strychnia has, however, been found in it, and there is no definite account of its origin. Prof. W. A. Hammond, from numerous experiments made with the drug, believed its action to be exerted mainly on the heart, paralyzing that organ. It was also thought to produce a paralysis of the sympathetic and motor nerves. Woorara is employed by the natives of South America to poison the heads of arrows. It exerts its peculiar effects by being introduced through wounds. When taken into the stomach it is often inert. The symptoms attending a wound poisoned with woorara are sudden stupor and insensibility, frothing at the mouth, rapid cessations of the respiratory movements and pulsations of the heart. Some writers say that the heart continues its action some moments after respiration has ceased.

*Treatment.*—When the poison enters a wound, the part should be sucked and excised, and a ligature placed around the limb between the wound and the heart. Brainard and Green discovered that a solution of iodine and iodide of potassium neutralized the poison, and recommend its application to the wound, and also its internal administration. Chlorine and bromine are also said to have a similar effect.

Artificial respiration has been tried on criminals poisoned by woorara, and has been followed by good results.

CALABAR BEAN.

. Calabar bean is a seed of the *Plysostigma venonosum*, a climbing plant of Calabar. It is used by the negroes of Africa as an ordeal-bean—the guilt or innocence of the individual being determined by its action on the system. If a dose is taken without subsequent unfavorable symptoms, the person is declared innocent. If the contrary, a verdict of guilty is announced.

Its action on animals is said to resemble that of woorara. It paralyzes the heart and motor nerves.

Poisonous doses in man produce vertigo, dimness of vision, great weakness, small, intermittent pulse, contraction of the pupil, insensibility, and death.

*Treatment.*—The stomach should be evacuated, stimulants administered internally, and the surface briskly rubbed. Hypodermic injections of strychnia might be tried. Strychnia exerts an entirely opposite effect on the spinal cord. Electricity is also worthy of a trial.

UPAS-TREE.

This tree grows in various parts of the East Indies. A resinous exudation, obtained by incisions in the bark, acts on the system as a virulent poison. Like woorara, it is principally employed by the natives to poison arrow-heads. The vapor of the tree at certain seasons of the year is said to cause eruptions on the skin.

When applied to a wound, or taken internally, it causes great muscular weakness, syncope, nausea, and vomiting, relaxed sphincters, thready, irregular pulse, and convulsions.

*Treatment.*—The remedies employed in poisoning by tobacco, or aconite, are applicable to these cases.

### SPINANTS.

### NUX-VOMICA (*Strychnia*).

Strychnia is derived from the seed of the *Strychnos nux-vomica* and the *Strychnos ignatia*, large trees of the East Indies and other Eastern countries. The seeds are embedded in the pulp of the fruit. They are circular in shape, three-quarters of an inch wide, about the thickness of a cent-piece, and are covered with delicate, yellowish-gray hairs. Strychnia exists in the seed, together with brucia and igasuria. The nux-vomica and its alkaloids possess the same action on the system, the only difference being in the rapidity with which their characteristic symptoms are manifested. Strychnia, which is the most powerful ingredient of the nut, or seed, is found in the shops in the form of a fine, white, crystalline powder, with an extremely bitter taste. Its bitterness is so marked that one part will give a taste to six hundred thousand parts of water (*U. S. Disp.*). Very small quantities suffice to produce a fatal result; one-tenth of a grain has killed a dog. There are instances recorded where half a grain has proved fatal to human beings. In exceptional cases recovery has taken place after the administration of four or five grains.

Strychnia acts specially on the spinal cord, but there is no good reason for supposing that it does not in a measure affect the brain. I have seen a certain amount of vertigo and rapid utterance follow its use.

There are several tests of the presence of this drug. In Mararchard's process, five or six drops of concentrated sulphuric acid, and one hundredth part of nitric acid, are mixed with the suspected solution; a little protoxide of lead is then added, and, if the strychnia is present, a blue color appears, which changes to violet, red, and finally to yellow.

If the strychnia is in solution in sulphuric acid, the addition of a bichromate-of-potash solution will give a violet hue. This test will detect the one million five hundred thousandth part of a grain (*U. S. Disp.*).

Poisonous doses of strychnia first produce an inability to remain in one position, and a tendency to perform every motion with great rapidity. The muscles seem to be beyond control of the will, and twitch unceasingly. There are some constriction in the throat, difficult respiration, and feeling of oppression about the chest. Violent muscular spasms then appear; they are tonic or continuous in character, resembling those occurring in tetanus. The muscles of the back are often affected more than those of the extremities, and as a result the body is bent like a bow, and rests on the head and heels (*opisthotonos*). During the paroxysm the jaws are tightly fixed, the face dark and congested from the accumulation of blood in the veins. Contraction of the muscles prevents expansion of the chest, and obstructs the blood going to the thorax, and hence the congestion. Intermissions in the severity of the paroxysms may occur; they last but a moment. Death takes place from the spasm of the muscles of respiration inducing asphyxia.

On *post-mortem* examination there are usually a dark color of the face, congestion of the brain, cord, and their membranes, and congestion of the lungs. The right side of the heart contains a large quantity of dark blood, and the left side is empty.

*Treatment.*—Chloroform taken in a liquid state or by inhalations should in all cases be tried. A relaxation of the spasms will at least prevent or retard the occurrence of asphyxia. Infusion of tobacco is recommended by some. It may be advantageously combined with chloroform; that is, the tobacco-infusion can be swallowed, or given by enema, while anæsthesia is procured by inhalation of chloroform. Aconite has been used in some cases with benefit. Thoral employs preparations of antimony as an antidote; it is given in emetic doses. Boudecker experimented upon dogs with chlorine-water and tartar-emetic, giving them alternately. He claims to have saved the animals from the poisonous effects of strychnia by this treatment.

It will be well in most instances to commence treatment by an emetic, in order to get rid of the poison remaining in the stomach. The infusion of tobacco, or sulphate of zinc, will answer this purpose. If the patient cannot swal-low the medicine, it can be given through the rectum.

A PECULIAR Spanish fly, called the *Cantharis vesicato-riu*, has long been employed in medicine as a vesicant and as a stimulant to the genito-urinary apparatus. There are several other varieties of cantharides found in the southern parts of this country, which possess properties analogous to the Spanish fly; they are, however, rarely employed for medicinal purposes.

Large doses of cantharides produce tenesmus at the neck of the bladder, inability to pass water, intense pain and scalding with the few drops of urine which are squeezed through (*strangury*), great pain throughout the alimentary canal, and thirst, with profuse vomiting and purging. The vomited matters and the stools contain blood. The extremities are cold. There are great prostration, a rapid pulse, sighing respiration, and a fetid odor to the breath.

A *post-mortem* examination shows signs of inflammation in the stomach and intestinal canal.

*Treatment.*—When the stomach and bowels have been emptied of their contents by emetics, cathartics, or the natural efforts of the patient, ten to thirty drops of liquor potassa largely diluted may be given every hour (*Mulack*), in conjunction with hot applications to the hypogastric regions. Small pieces of ice may be swallowed with benefit. Thale recommends animal charcoal as an antidote; a teaspoonful of this substance mixed with a little water may be given at a dose.

### OIL OF SAVIN.

The tops and leaves of *Juniperus sabina,* or red cedar, furnish a volatile oil which possesses marked irritant properties. The oil is employed in medicine as a stimulant to the secretions, and as an emmenagogue. Its action on the uterus is denied by some authorities. It is commonly administered by quacks and others to produce abortion. These cases not infrequently terminate fatally.

A decoction and infusion of the tops and leaves are also used for a similar purpose.

An overdose produces strangury, sharp pains in the bowels, hot skin, rapid pulse, violent vomiting, and sometimes purging. The vomited matters are often of a green color. Great prostration comes on rapidly, and usually ends in death.

The *post-mortem* appearances are the same as those observed in poisoning by cantharides.

*Treatment.*—Warm fomentations over the epigastrium and hypodermic injections of morphia may be tried with benefit. The patient should be fed through the rectum if possible. Nothing but ice should be allowed in the stomach until the subsidence of the inflammation.

### CROTON OIL.

Is a product of the seeds obtained from the *Croton tiglium,* a small tree of Hindostan. It is a drastic hydragogue cathartic, acting efficiently in from a half to one hour after its administration. Applied externally it produces a pustular eruption. In large doses it excites inflammation of the œsophagus, stomach, and intestines, and gives rise to vomiting, purging, and rapid prostration.

*Treatment.*—Empty the stomach thoroughly, and treat the resulting inflammation in the usual manner. Stimulants diluted with iced milk should also be used, to sustain the strength of the patient.

### COLCHICUM.

The tinctures and decoctions of this drug are not infrequently taken in poisonous doses by careless persons. Three drachms of the wine of the seeds have caused death. The poisonous effects are manifested by violent vomiting and purging, great pain, and collapse.

The treatment is the same as for the preceding varieties.

### VERATRIA.

This alkaloid is obtained from the seeds of *Veratrum sabadilla* and other plants. It is found in the shops, in the form of a grayish-white powder. The taste is bitter. It gives a red color with sulphuric acid, and a yellow color with nitric acid.

Veratria is a powerful poison in doses of four or five grains. Half a grain has proved fatal to a child.

The symptoms of poisoning are vomiting and purging, pain in the epigastrium, rapid respiration, small, quick pulse, and spasmodic movements of the muscles, resembling those which occur in tetanus.

The antidotes are vinegar, vegetable astringents, Lugol's solution, and stimulants.

Black and white hellebore, all the drastic cathartics, turpentine, etc., are irritant poisons in large doses. They present similar symptoms to those irritants previously mentioned, and require the same treatment.

# CHAPTER XXII.

ARSENIC.

EVERY preparation of arsenic acts as an irritant poison. Among the most common varieties are arsenious acid, arsenite of copper (Scheele's green); yellow sulphuret of arsenic (orpiment); and red arsenic, or realgar. Arsenious acid and Scheele's green are most frequently employed for purposes of murder or suicide.

Metallic arsenic is made by heating an oxide of arsenic with charcoal.

Arsenious acid ($AsO_3$) is obtained during the sublimation of the arseniuret of cobalt and iron. It usually exists in the shops as a fine white powder. If the sublimation has been slow, it will take the form of brilliant octahedral crystals (*Taylor*). It combines with many of the alkalies, as soda, ammonia, or potash, to form salts. The well-known Fowler's solution is a liquid preparation of the arsenite of potash.

Scheele's green is applied to a variety of purposes. It is the principal ingredient in the coloring matter of green wall-paper, artificial flowers, candy and paper boxes, etc. Nearly all the bright-green colors of household furniture, paper, and "knick-knacks," are made by this poison. This indiscriminate and unguarded use has resulted in serious impairment of health and loss of life. Inhalations of the microscopical particles, which arise from the green surface of room-paper, may induce all the poisonous effects of arsenic. Cases are not rare where this has occurred.

Realgar and orpiment are much used also as coloring matters, but less extensively than arsenite of copper.

Arsenious acid ($AsO_3$) is a very powerful poison, but loss of life from its administration is exceedingly rare.

Arsenious acid kills in from three to forty-eight hours. The length of time varies with the dose, the condition of the stomach, and age of the patient. Christoson gives the smallest fatal doses of the preparation as thirty grains of the powder, and four grains in solution. Taylor relates a case where two or three grains in powder proved fatal.

*Tests.*—Ammonia nitrate of silver, added to a solution of arsenious acid, throws down a yellow precipitate, which is the arsenite of silver. Ammonia sulphate of copper gives a green precipitate of arsenite of copper.

Marsh's test is the most reliable. It consists in adding sulphuric acid and zinc to the arsenical solution, and forming arsenuretted hydrogen. The gas, as it passes out through the tube, is set on fire. The presence of arsenic is known by the garlicky odor, and by the blue color of the flame. In addition, if a porcelain slate is held near the flame, a black ring of metallic arsenic is deposited, and on the outside of this ring a whitish film of arsenious acid appears. To determine whether the deposit is arsenic or antimony, the plate is subjected to a high temperature, and, if arsenic is present, the substance is immediately volatilized; if antimony, it will remain.

Riensch's test consists in boiling slips of copper in an acidulated solution of the suspected liquid. The mixture is heated to the boiling-point, and a slip of copper dipped in it for five or ten minutes. If arsenic is present, it will be deposited on the copper, and will appear of a dark-gray color. If the material thus obtained is heated in a tube, the metallic arsenic is changed into arsenious acid, which is recognized by its peculiar bright octahedral crystals.

Scheele's green and other preparations of arsenic are distinguished by the same reagents. In all cases the arsenic may be reduced to arsenious acid by heat, while the latter can be recognized by its crystals.

Small and repeated doses of arsenic may produce slow poisoning. The constitutional effects of the drug administered in this manner are recognized by a pale, waxy look on the face, œdema of the eyelids and sometimes of the extremities, loss of appetite, pain in the stomach, nausea, or vomiting, eruptions on the cutaneous surface, feeble pulse, and great weakness. In some cases the urine is loaded with albumen. If the drug be continued, death soon ensues.

When large doses of arsenic are taken, there is pain in the epigastric region, which rapidly increases, and is aggravated by pressure. There are nausea and vomiting. At first, the vomited matter consists of the contents of the stomach, with particles of arsenic intermixed. Subsequently, they contain blood and thick mucus. Purging usually follows the vomiting, in about half an hour after the prominent symptoms are developed. There are sometimes soreness and constriction about the throat. The respiration becomes entirely thoracic, and the movements are short and

16

rapid. The pulse is quick, small, and intermittent. Death may be preceded by coma and convulsions.

In poisoning from corrosive sublimate, the symptoms are developed more rapidly than in arsenical poisoning. In the former there is greater pain in the throat, and in the course of the œsophagus, and the tongue, fauces, and throat, present a white appearance. These signs suffice to distinguish the two forms.

After death from arsenic, the mucous membrane of the stomach is congested, thickened, and softened. There is more or less redness over the whole organ, but marked in the most dependent portions. Collections of mucus, mixed with blood and arsenic, are found in isolated patches in different parts of the stomach. Arsenic does not act as a corrosive poison; it never produces ulceration of the mucous membrane.

*Treatment.*—The antidote for arsenious acid is the hydrated sesquioxide of iron. It is prepared by adding aqua ammonia, soda, or potash, to a solution of the persulphate of iron. When the alkali is added, a reddish-brown powder forms, which is administered *ad libitum* both to adults and children. The iron combines with the arsenic, and the insoluble subarseniate of the protoxide of iron is thrown down (*U. S. Disp.*). Preceding the administration of the antidote, the stomach should be thoroughly emptied with the stomach-pump, or by emetics of sulphate of zinc, mustard, or ipecac., assisted by copious draughts of warm water.

Preparations of magnesia are recommended as antidotes. Lime-water, mixed with oil, and mucilaginous drinks, may be given also.

The antidote for the salts of arsenic is the subacetate of the protoxide of iron (*Duflos*).

Fewtrell recommends the administration of a mixture of chalk and castor-oil, made into a thick paste.

When the stomach is cleansed and the antidotes given, the treatment should be directed to allay pain, and relieve the gastric inflammation, by hypodermic injections of morphia, internal administration of ice, and blisters to the epigastrium.

CORROSIVE SUBLIMATE (*Bichloride of Mercury*).

Mercury in the metallic state is inert. When taken internally it passes through the bowels with scarcely any change. An extremely small quantity may be oxidized, but not sufficient to affect the system. Many of the combinations of mercury act as corrosive and irritant poisons. The most deadly is corrosive sublimate. This substance, according to American authorities, consists of two atoms of chlorine united to one of mercury. The British Pharmacopœia, however, makes it a protochloride, consisting of equal parts of chlorine and mercury. The bichloride is made by subliming sulphuric acid and mercury together, and then adding chloride of sodium. It occurs in small white or transparent crystals, and is exceedingly soluble.

*Tests.*—Iodide of potassium gives a scarlet-colored precipitate of the biniodide of mercury. Ammonia throws down a white precipitate of ammoniated mercury. Lime-water gives a yellow precipitate of the hydrated deutoxide of mercury. A black precipitate is formed by sulphuretted hydrogen. If a piece of zinc and gold wire be dipped in the suspected solution, which has been slightly acidulated, a grayish deposit of mercury will take place on the metal.

In small doses, continued, it produces ptyalism and other characteristic effects of mercurial preparations. The patient's gums become red, tender, swollen and ulcerated; saliva is poured out in excessive quantities. There is a strong metallic taste in the mouth, and the breath has a fetid odor. A blue line, in some cases, may be noticed around the edge of the gums. The teeth loosen, and the throat becomes sore and inflamed. The blood loses its plasticity, and the red globules are diminished. If allowed to proceed without treatment, these symptoms are intensified; necrosis of bone and ulceration of the integument are added, and the patient dies from exhaustion.

Corrosive sublimate has been known to destroy life in doses of three grains (*Taylor*). Usually it takes from ten grains to a drachm. In a few cases much larger doses have been recovered from.

The symptoms produced by poisonous doses are those common to many corrosive poisons. A burning pain is felt along the œsophagus and in the stomach, a few moments after the drug is swallowed. This is followed by vomiting and purging of slimy mucus, marked with blood. Portions of mucous membrane have been thrown up with the evacuations. The mouth and throat have a white appearance, and a strong metallic taste is experienced. There are thirst, difficulty in swallowing, a feeling of oppression on the chest, and difficulty in breathing. The pain in the stomach increases in intensity, the pulse becomes small and thready, extremities cold; great prostration comes on, which is soon followed by death

The mucous membrane lining the œsophagus and stomach present after death a slate-gray appearance. The membrane is softened, and may be ulcerated. Extravasations of blood are found beneath it, and occasionally on the surface. If a piece of the membrane is taken up with a forceps, it is easily separated. There are also redness and tumefaction, particularly marked in the great *cul-de-sac* of the stomach.

*Treatment.* — When profuse salivation arises from medicinal doses of corrosive sublimate, or other preparations of mercury, iodide of potassium is given as an antidote in conjunction with chlorate of potash. A solution of the latter makes an efficient wash for the ulcerated mouth. Carbonic acid, in the proportion of one drachm to four ounces of water, is an excellent application for the same part.

When poisonous doses of the bichloride have been taken, the stomach should be emptied rapidly and completely with emetics or the stomach-pump. The common antidote, *albumen*, may then be administered, in the form of white of egg, or the *gluten* of bread. The egg should be beaten up with a large quantity of water before it is given. Milk may also be administered in large quantities. The casein it contains, as well as the albumen of the egg, forms an insoluble compound with the mercury. Small rolls of zinc and gold foil have been recommended as antidotes.

The subsequent inflammation should be treated in the same manner as that arising from arsenical poisoning (*see* Arsenic). .

*Calomel* is the true protochloride of mercury. It acts sometimes as an irritant poison, but there are few cases of destruction of life from its use. It is recognized by its extreme insolubility. The bile is the only fluid in the body which exerts a solvent action upon it, and that only in very small proportions. Potash and ammonia give a black precipitate; lime-water gives also a black precipitate.

### COPPER.

The preparations of copper in common use are the sulphate (*blue vitriol*) and subacetate (*verdigris*). The sulphate of copper is employed medicinally, internally, as an emetic, and externally as an escharotic. Verdigris possesses similar properties, but is little used.

Chronic poisoning from copper may be induced by working in alloys of that metal, inhaling copper-dust, or eating from utensils lined with that metal.

All the soluble preparations of copper are corrosive poisons, and the effects on the system similarly manifested. The quantity of sulphate of copper which will destroy life is subject to great variation. Being a powerful emetic, the poison is rapidly thrown from the stomach, and the danger lessened. Nearly an ounce of the poison has been taken and recovered from, while in another instance one drachm has been known to destroy life.

*Tests.*—Ammonia, potash, and soda, give a bluish-white precipitate. Ferrocyanide of potassium gives a claret red precipitate (*Taylor*).

When the system becomes slowly impregnated with copper, there are a rapid loss of flesh and strength, nausea, tendency to diarrhœa, griping abdominal pains, tympanitis, muscular tremors, retraction of the gums, with a purple line around the edge (*Corrigan*), a dry cough, paralysis, dysenteric discharges from the bowels, and great prostration.

In acute poisoning there are intense griping pains in the abdomen, profuse greenish-colored discharges from the stomach and bowels, metallic taste in the mouth, anxious facies, vertigo, headache, dimness of vision, muscular tremors, a rapid, small pulse, paralysis, and sometimes convulsions.

After death, the mucous membrane of the œsophagus, stomach, and intestines, is reddened and softened. Ulceration and erosion in patches are found in different parts of the canal.

*Treatment.*—Ferrocyanide of potassium is recommended as an antidote by Schræder. Milk and honey, or white of egg, and milk in copious draughts,. are often serviceable. Albumen in any form, or sugar, is considered, by many, an efficient antidote.

The resulting gastro-enteritis is treated as in the preceding cases.

### LEAD.

Every soluble salt of lead possesses poisonous properties. The carbonate and oxide are more frequently the active agents in chronic poisoning than any other preparations. The acetate (sugar of lead), and the solution of the subacetate (Goulard's extract), occasionally exert a deleterious effect on the system, when given in ordinary medicinal doses. The carbonate of lead (white lead) is more severe in its action than the other salts. Usually a very large quantity of lead is necessary to destroy life.

Chronic poisoning is of frequent occurrence, from using hair-dyes, drinking beer or water which flows through lead pipes, constant handling of the thin foil covering chewing-tobacco, manufacturing or mixing white lead. It is sometimes produced by wearing Brussels lace, the material of which owes its white color to carbonate of lead.

*Tests.*—Sulphuric acid throws down a white precipitate. Iodide of potassium gives a yellow, and sulphuretted hydrogen a black precipitate.

The symptoms of poisoning by lead appear gradually. There are, at first, colicky pains in the abdomen, and constipation. The attack of colic (*colica pictonum*) may be very severe, or so slight as scarcely to demand attention. It is paroxysmal in character. The bowels are constipated. A blue line appears around the edge of the gums. There are " thumb-drop " and wrist-drop," from paralysis of the extensor muscles. The right rectus abdominalis is said to be the first muscle affected by the paralysis. The retraction of the abdomen witnessed in these cases is due to paralysis of those muscles. Paraplegia and hemiplegia exist in rare cases. Loss of flesh and strength, and muscular tremors, are also present.

When very large doses of lead are taken, there are thirst, dryness of the fauces, burning sensation in the throat, constipation, and intense colicky pains in the abdomen. If the bowels are moved, the fæces will be found to possess a dark color due to the change of the lead into the sulphuret in the intestinal canal (the same color is also observed after the administration of iron; the iron is changed into the sulphuret). Vomiting is sometimes present; there are difficult respiration and oppression over the præcordia. Paralysis and coma precede death.

On *post-mortem* examination there is usually found abrasion of the mucous membrane of the stomach and intestines, with redness and congestion in isolated patches;

also, a grayish-white color in certain portions, from the mixing of the mucus with the lead.

*Treatment.*—In chronic poisoning, iodide of potassium is considered the best eliminative. It joins with the lead in the system to form a soluble iodide of lead, which is carried out through the different emunctories. Sulphuric acid is sometimes administered for the same purpose. The patient should entirely change his habits, take active exercise in the open air, eat nourishing food, and keep regular hours. Quinine is a useful tonic in these cases. The paralyzed limbs may be treated by frequent bathing in cold water and by friction.

In acute poisoning from lead, the stomach should first be emptied by emetics, or with the stomach-pump. Strong solutions of Epsom salts (sulphate of magnesia), or Glauber's salts (sulphate of soda), may then be given in large quantities, as antidotes. If the bowels do not move, castor-oil should be given until free evacuations are produced. Animal charcoal is given by some. Albumen and milk may be used after or before the administration of the salts of magnesia or soda. These are not unfrequently employed alone. Taylor advises a mixture of vinegar and sulphate of magnesia as an antidote for poisoning by the carbonate of lead.

### TARTARIZED ANTIMONY.

This substance is prepared by adding an ounce of the oxide of antimony, and one ounce of bitartrate of potash to eighteen ounces of water, and then boiling for one hour. Tartarized antimony is used in medicine as an emetic, sedative, alterative, diaphoretic, and expectorant. In large doses is an irritant poison.  The ordinary dose for an adult, as an emetic, is from one to two grains; with young persons very small doses will often produce dangerous effects.   Three-quarters of a grain has been known to destroy life in a child (*Wilton*).  Ten grains is the smallest recorded fatal dose in an adult.   Although antimony is capable of producing rapid, violent constitutional disturbances, yet remedial efforts are generally followed by recovery.   It is not apt to prove fatal, with proper care.

*Tests.*—Nitric acid throws down a white precipitate, which is soluble in tartaric acid.   Sulphuretted hydrogen gives a characteristic red color to a solution of antimony, and, if muriatic acid is added to the precipitate, it is dissolved.   If the solution is then added to water, a white precipitate appears.

Chronic poisoning by tartarized antimony is distinguished by gradual exhaustion, nausea, and vomiting, pain in the epigastrium, a small, feeble pulse, pallid surface, and cold, clammy extremities, sunken eyes, anxious expression of countenance, and metallic taste in the mouth.

In large quantities the drug produces in a few moments profuse bilious vomiting, and the matter vomited is soon mixed with blood.   Portions of mucous membrane, of a grayish-white or dark-brown color, may come away in small pieces (*Taylor*).   Diarrhœa is present if much of the poison has been swallowed.   Signs of collapse are apparent: the skin becomes cold and bathed in a clammy perspiration, the pulse is feeble and rapid, and respiration sighing.   A pustular eruption has been observed on the skin in some cases.   Before death, the patient sinks into a deep coma.

A *post-mortem* examination shows signs of inflammation

in the throat, stomach, and intestines. Patches of mucous membrane, softened and easily detached and broken down, are found in the throat and stomach, and occasionally in the small intestines. Peritonitis is found in a small proportion of cases. The lungs are congested.

*Treatment.*—Large quantities of warm water should be given, to promote the complete evacuation of the stomach. Strong infusions of green tea may be taken at the same time or subsequently; various vegetable astringents, as tannic acid, etc., are also used as antidotes. Attempts should be made to counteract the collapse by hot bottles and blankets applied to the surface, and by friction of the extremities.

### ZINC.

Sulphate of zinc, or *white vitriol*, and chloride of zinc, are energetic poisons; the former is an irritant, the latter a corrosive poison. The sulphate is employed in medicine as an astringent, nervine, and emetic. Its dose, as an emetic, is from ten to twenty grains. The chloride of zinc in solution is a valuable disinfectant.

The tests for zinc are ammonia, ferrocyanide of potassium, and sulphuretted hydrogen, all of which give a white precipitate.

In poisoning from white vitriol, there are nausea and vomiting, pain in the abdomen, followed by all the signs of collapse. When the chloride is the poisoning agent, the pain and collapse are greater; there are lividity of the surface, vertigo, and dimness of vision. In the evacuations from the stomach, shreds of mucous membrane are found.

The stomach, after death, is dark-colored; the mucous membrane thickened, congested, and perhaps ulcerated.

*Treatment.* — White of egg, beaten up with milk and water, followed by infusions of astringent medicines, is the chief remedy for poisoning from the sulphate.

In poisoning from the chloride, emetics should first be given; the albumen in milk can be administered when the stomach has been emptied.

### NITRATE OF SILVER.

This substance is a corrosive poison. It has powerful escharotic properties, due to its affinity for the albumen of the tissues.

In poisonous doses, it produces intense pain, vomiting, and purging. Mucus, blood, and shreds of mucous membrane, are found in the excavations. If these are allowed to stand, they become dark from exposure to air.

Common salt (chloride of sodium) throws down a white precipitate with solutions of nitrate of silver, and it is also given as an antidote. Mucilaginous drinks should be administered *ad libitum.*

### PHOSPHORUS.

Phosphorus is largely employed in the manufacture of lucifer matches. It is seldom used for medicinal purposes. Children are frequently poisoned by sucking the ends of matches, or drinking water in which they have been soaked. In match-factories, chronic poisoning from inhalation of phosphorus-vapor is of common occurrence. The symptoms of acute poisoning from phosphorus are peculiar in not developing for some hours after the poison has been taken. A small amount, one-tenth of a grain, has caused death.

Phosphorus is recognized by its peculiar odor, and its luminous appearance in the dark.

Chronic poisoning usually manifests itself first by ordinary dyspeptic symptoms; such as loss of appetite, feeling of weight and heat in the epigastrium, and by prostration. There are also nausea, diarrhœa, restlessness, inability to sleep, pains in the bones, and febrile excitement, which is worse toward night. If the exposure to the poisonous vapor have been of long duration, necrosis of the lower jaw, low grades of inflammation in various parts, and congestion of the lungs, will be found, in addition to the other symptoms.

In acute poisoning there are vomiting and purging of a greenish-colored substance, which soon becomes mixed with blood and mucus. The ejections and breath have a garlicky odor. If brought to a dark place, they exhibit a peculiar luminous appearance. There are intense pain in the abdomen, and tympanites. The face is anxious, skin cold, and the pulse is rapid and small. A fatal termination does not, usually, take place until a day or two has elapsed from the commencement of the symptoms, and in some cases life has been prolonged for a week.

After death the stomach presents signs of gangrenous inflammation. The mucous membrane is intensely red, and easily detached and broken down. There may be perforations in the wall of the intestines, passing into the peritoneal cavity. Congestion of the brain and serous effusion into the ventricles are also present. The viscera have a garlicky odor, and, when exposed in a dark place, become luminous.

*Treatment.*—Phosphorus has no direct antidote. Taylor recommends hydrated magnesia, and the free use of demulcent drinks, and albumen.

# CHAPTER XXIII.

*CORROSIVE ACIDS.*

### OXALIC ACID.

THIS substance exists in combination with potash in sorrel, with lime in rhubarb; it is found also in a free state in the chick-pea. It is made by the action of nitric acid on sugar; or upon rice, gum, starch, etc. Chemically, it is composed of one atom of carbonic oxide, and one atom of carbonic acid, making its formula $C_2O_3$.

The crystals of oxalic acid are sometimes mistaken for those of Epsom salts. The crystals of the former are distinguished by having a sour taste, and by being clearer and more transparent than those of Epsom salts. The crystals of the latter have a bitter taste.

Oxalic acid is a deadly poison, acting with great rapidity, and causing death in from five minutes to half an hour.

*Tests.*—Chloride of calcium gives a white precipitate of oxalate of lime; sulphate of copper, a bluish-white precipitate of oxalate of copper; and nitrate of silver, a white precipitate of oxalate of silver (*Wood & Bache*).

Oxalic acid, when given in a concentrated form, produces pain in the throat, œsophagus, and stomach. The vomiting is associated with violent retching. There are rapid prostration, syncope, and death.

If largely diluted, its corrosive action is decreased, and the symptoms are not so violent. There are less pain and vomiting, but stupor and prostration are more distinctly marked. Death may result from paralysis of the heart. Christosin states that the mucous membrane after death has a scalded appearance, that dark-colored spots are found scattered through the whole canal, and that the membrane is entirely destroyed in some parts, leaving the muscular coat bare.

*Treatment.*—Emetics should be given and followed immediately by the antidotes. Lime or magnesia should be administered in large quantities in water. The lime is usually employed in the form of the carbonate (common chalk). If this cannot be had, the ceiling of the room may be scraped with a shovel or other available instrument, and the substance thus obtained given in the manner prescribed. Lime and magnesia form insoluble salts by combining in the stomach with the oxalic acid.

### SULPHURIC ACID.

There are three varieties of this acid, viz., the anhydrous, $SO^3$; commercial, $SO_3 + Ho$; and the fuming oil of Nordhausin, $SO_3Ho + SO_3$.

The commercial sulphuric acid, which is the variety generally employed for medicinal purposes, is made by burning sulphur and nitrate of potash together in a leaden chamber containing water. It is a powerful corrosive poison, destroying organic tissues when brought in contact with them. It has a powerful affinity for water, and its caustic effect is due to the abstraction of that substance from the tissues. It makes a red stain on black cloth.

*Tests.*—Chloride of barium throws down a white precipitate.

In poisoning from sulphuric acid, the pain is most intense in the mouth, throat, œsophagus, and stomach. There are great pain on pressure, vomiting of black putrid matter, dyspnœa, small, feeble pulse, anxious expression of countenance, cold extremities, restlessness, and sometimes convulsions.

*Treatment.*—The poison may be neutralized by magnesia, or carbonate of soda, administered in solution, thick soap-suds, and mucilaginous drinks. Unless these remedies can be given directly after the poison has been swallowed, there is little chance of saving the life of the patient.

### NITRIC ACID.

Nitric acid is made by the action of sulphuric acid on nitrate of potash. It is a powerful corrosive poison. In medicine it is employed as a tonic, astringent, and antispasmodic. The vapor of nitric acid is reputed a good disinfectant. Inhaling the vapor in a concentrated form has produced death. One to two drachms of the liquid have been known to destroy life.

*Tests.*—A solution of morphia added to nitric acid gives a red color, which afterward changes to a yellow. If the acid is boiled in water containing copper filings, red fumes of nitrous acid are given off. When applied to clothing it gives a yellow stain.

The symptoms of poisoning are violent pain, extending

from the mouth to the epigastrium, vomiting of yellowish and greenish-black material, and the emission of fetid gas, tympanitis, urgent dyspnœa, small, rapid pulse, and collapse. Constipation is usually present. The enamel of the teeth will be found partially destroyed; the tongue, throat, and fauces, of a yellowish-brown color, and very much swollen.

If poisoning have resulted from inhalations of the vapor, there will be great pain, difficulty in respiration, and the patient may die asphyxiated from effusion under and into the mucous membrane of the larynx.

After death, the mucous membrane of all parts of the alimentary canal which came in contact with the poison is deeply corroded; in some parts there are yellowish-brown stains, in other parts extensive redness. The mucous membrane is readily broken down; in many cases there is congestion of the lungs and larynx.

*Treatment.* — Magnesia, olive-oil, and mucilaginous drinks, should be given in large quantities.

### MURIATIC ACID.

This acid is made by the action of sulphuric acid on chloride of sodium. It is sometimes called *spirit of salt.* Cases of poisoning by it are rare.

*Tests.*—If the acid is boiled with black oxide of manganese, chlorine is evolved, which is recognized by its odor and its bleaching properties. If a rod is dipped in the acid and held near ammonia, a white vapor of the hydrochlorate of ammonia is formed. Nitrate of silver throws down a white precipitate of chloride of silver.

The symptoms following large doses resemble those
17

produced by the other corrosive acids. They are, however, developed more slowly; life is not so soon destroyed, and white vapors may be emitted from the mouth.

### CARBOLIC ACID,

Sometimes called oxide of phenyl, or phenylic acid, is much employed at the present day as a disinfectant. It is obtained by the distillation of coal-tar. Very few cases of poisoning by it have yet occurred.

A concentrated solution taken internally excites violent gastro-enteritis, and destroys life in a few hours.

After death, the mucous membrane of the throat and stomach is intensely congested, and in small sections softened and corroded.

The treatment consists in evacuating the stomach, and giving large quantities of magnesia, mucilaginous drinks, etc.

# CHAPTER XXIV.

### SALTS OF POTASH.

CARBONATE of potash (pearlash) acts as a corrosive poison when administered in a concentrated form. It gives a yellowish-white precipitate with nitrate of silver. The symptoms following its administration are intense pain in the throat and stomach, pain on pressure over the abdomen, vomiting of dark materials, which consist of mucus, blood, and shreds of the lining membrane. Diarrhœa occurs in all cases. On examination, the mouth and throat are found of a dark-red color, and very much swollen. This condition seriously interferes with deglutition. The pulse is small, rapid, and weak, and the countenance anxious.

After death the mucous membrane of the throat and stomach is of a dark-brown color, softened, and in some portions destroyed.

*Treatment.*—Taylor advises the use of citric or acetic acid, lemon or orange juice. Oil in large quantities, and mucilaginous drinks, are efficient remedies.

*Hydrated oxide of potassium,* or caustic potash, is distinguished from the carbonate by giving a brown precipitate with nitrate of silver.

The symptoms produced by poisoning with this drug are

similar to those which occur after administration of the carbonate, and a like treatment is necessary.

Binoxalate of potash, sometimes called essential salt of lemon, is an active poison, resembling oxalic acid in its effects on the system. It is sometimes mistaken for cream of tartar. The latter, however, is not precipitated from its solution by the sulphate of lime, while the former is. Ink-stains are removed by the binoxalate, which furnishes another distinguishing point.

The symptoms of poisoning are violent vomiting and purging, pain in the stomach, difficult deglutition, and sighing respiration, small, rapid pulse, cold extremities, great prostration, and muscular spasms.

*Treatment* consists in the administration of lime, magnesia, and mucilaginous drinks.

### NITRATE OF POTASH,

Usually known as *saltpetre*, is employed medicinally as an antiseptic, diuretic, refrigerant, diaphoretic, and sedative. In doses of from three drachms to an ounce it acts as a corrosive poison.

In these doses it causes vomiting and purging of blood and mucus, violent pain in the abdomen; there are feeble pulse, rapid prostration, insensibility, and death.

*Treatment.*—The stomach should be emptied by emetics, and mucilaginous drinks should be freely administered; opium should be given to relieve pain.

There is no antidote for the poison. The salts of soda correspond with the salts of potash in their peculiar poisonous action, and in the treatment.

## AMMONIA.

Strong solutions of ammonia, carbonate and muriate of ammonia, act as corrosive poisons. The vapor of ammonia, when inhaled in large quantities, excites inflammation of the mouth, fauces, and air-passages, and may produce asphyxia. Solutions of the carbonate (*sal-volatile*), or of gaseous ammonia, produce violent inflammation in the œsophagus and stomach, and corrode the mucous membrane. The carbonate is said to be more violent in its action than the other preparations.

These substances are recognized by their peculiar penetrating odor.

The symptoms of poisoning are nausea, and vomiting of mucus, mixed with blood and shreds of mucous membrane, pain in the throat and epigastrium. Perforations of the stomach sometimes take place, and are followed by peritonitis. There is great difficulty in swallowing and breathing. The mouth is tender and swollen, the face is anxious, the pulse rapid and feeble, and the extremities cold.

After death the blood is found more fluid than in other cases of poisoning; there are extravasations of blood in the stomach and intestines, and congestion, softening, and erosion of the mucous membrane.

*Treatment.*—Vinegar, acetic acid, diluted milk, and mucilaginous drinks, are usually given; opium is necessary to relieve pain.

# APPENDIX.

## PAGE 20.

I have used the aspirator in thirteen cases with success. But in my recent cases I have used Colin's instrument, instead of the aspirator, because of its universal applicability, and also on account of its simplicity and safety. It consists of a reservoir or basin for the blood, to which is attached a syringe—nothing more than an ordinary syringe—working as any syringe would. To this is attached a tube, ending in a canula, which fits accurately a canula previously inserted into the vein of the patient. The entrance to the tube of exit is guarded by an aluminum ball-valve which excludes the air, and permits only the blood to enter. Thus one of the principal dangers in transfusion is avoided.

Though Colin's instrument is intended to throw in blood in its natural state, undefibrinated, I have, in the majority of my cases, in order to make the operation absolutely safe, defibrinated the blood, and added it to a solution of ammonia and water, already contained in the basin of the instrument.

## PAGE 38.

I have lately devised a tourniquet on the principle of the "safety-pin," which obviates danger from hæmorrhage during the operation, and enables the surgeon to control all the bleeding parts without difficulty. The instrument was first employed to control bleeding from the lingual artery during amputation of the tongue. It can be utilized in a great variety of operations.

## PAGE 54.

Dr. Lewis, of Philadelphia, has devised an ingenious method of suppressing hæmorrhage from the intercostal arteries. He passes the handle of an ordinary door-key through the wound, turns the end at right angles to the rib, then by very slight pressure compresses the artery.

# INDEX.

THE END.

# REASONS WHY PHYSICIANS SHOULD

## SUBSCRIBE FOR THE

# ℿew York ℿedical Journal,

### EDITED BY FRANK P. FOSTER, M. D.

BECAUSE: It is the *LEADING JOURNAL* of America, and contains more reading-matter than any other journal of its class.

BECAUSE: It is the exponent of the most advanced scientific medical thought.

BECAUSE: Its contributors are among the most learned medical men of this country.

BECAUSE: Its "Original Articles" are the results of scientific observation and research, and are of infinite practical value to the practitioner.

BECAUSE: The "Reports on the Progress of Medicine," which are published from time to time, contain the most recent discoveries in the various departments of medicine, and are written by practitioners especially qualified for the purpose.

BECAUSE: The column devoted in each number to "Therapeutical Notes" contains a *résumé* of the practical application of the most recent therapeutic novelties.

BECAUSE: The Society Proceedings, of which each number contains one or more, are the reports of the practical experience of prominent physicians who thus give to the profession the results of certain modes of treatment in given cases.

BECAUSE: The Editorial Columns are controlled only by the desire to promote the welfare, honor, and advancement of the science of medicine, as viewed from a standpoint looking to the best interests of the profession.

BECAUSE: Nothing is admitted to its columns that has not some bearing on medicine, or is not possessed of some practical value.

BECAUSE: It is published solely in the interests of medicine, and for the up-holding of the elevated position occupied by the profession of America.

**The volumes begin with January and July of each year. Subscriptions must be arranged to expire with the volume.**

### Subscription price, $5.00 per Annum.

New York: D. APPLETON & CO., 1, 8, & 5 Bond Street.

# JOURNAL OF
# CUTANEOUS AND GENITO-URINARY
# DISEASES.

*Edited by* { Prince A. Morrow, A. M., M. D., and
John A. Fordyce, M. D.

## PUBLISHED MONTHLY.

WITH. the number for January, 1889, this Journal enters upon the seventh year of its publication. The history of the Journal has been one of progression, and, under the present editorial management, there can be no doubt that it will preserve and increase the reputation already established.

Devoted to the diseases indicated in its title, the Journal will be contributed to by the most eminent dermatologists and syphilographers in this country. Whenever the subject requires illustration, wood-cuts or chromo-lithographs will be employed.

Letters from Europe, one or more of which will appear in each issue of the Journal, will keep the reader informed of the advances in this department of medicine at the great medical centers—Vienna, Berlin, and Paris.

A feature of the Journal will be the publication of abstracts of translations of notable papers and selections from foreign journals.

Due prominence will be given to Society Transactions, including papers read and the discussions had thereon, so far as they have a bearing upon the subjects to which the pages of the Journal are devoted.

Both the editors and the publishers will put forth every effort to make the Journal instructive, attractive, and a representative one of its class; and they feel assured that every practitioner, whose work brings him in contact with cutaneous or genito-urinary diseases, will find it of great value and assistance to him.

### Subscription price, $2.50 per Annum.

Subscriptions should be arranged to expire with either June or December number.

New York: D. APPLETON & CO., 1, 8, & 5 Bond Street

ESTABLISHED BY *EDWARD L. YOUMANS.*

# THE POPULAR SCIENCE MONTHLY,

### EDITED BY WILLIAM JAY YOUMANS,

Well known as a trustworthy medium for the spread of scientific truth in popular form, is filled with articles of interest to everybody, by the ablest writers of the time. Its range of topics, which is widening with the advance of science, includes—

PREVENTION OF DISEASE AND IMPROVEMENT OF THE RACE.
AGRICULTURAL AND FOOD PRODUCTS.
SOCIAL AND DOMESTIC ECONOMY.
POLITICAL SCIENCE, OR THE CONDUCT OF GOVERNMENT.
SCIENTIFIC ETHICS; MENTAL SCIENCE AND EDUCATION.
MAN'S ORIGIN AND DEVELOPMENT.
RELATIONS OF SCIENCE AND RELIGION.
THE INDUSTRIAL ARTS.
NATURAL HISTORY; DISCOVERY; EXPLORATION, ETC.

With other illustrations, each number contains a finely engraved PORTRAIT of some eminent scientist, with a BIOGRAPHICAL SKETCH. Among its recent contributors are :

| | |
|---|---|
| W. A. HAMMOND, M.D., | BENJ. WARD RICHARDSON, M.D., |
| HERBERT SPENCER, | ANDREW D. WHITE, |
| DAVID A. WELLS, | F. W. CLARKE, |
| T. H. HUXLEY, | HORATIO HALE, |
| SIR JOHN LUBBOCK, | EDWARD S. MORSE, |
| EDWARD ATKINSON, | J. S. NEWBERRY, |
| T. D. CROTHERS, M.D., | WALTER B. PLATT, M.D., |
| W. K. BROOKS, | EUGENE L. RICHARDS, |
| E. D. COPE, | THOMAS HILL, |
| DAVID STARR JORDAN, | N. S. SHALER, |
| T. MITCHELL PRUDDEN, M.D., | D. G. THOMPSON, |
| JOSEPH LE CONTE, | AMBROSE L. RANNEY, M.D., |
| APPLETON MORGAN, | GRANT ALLEN, |
| FELIX L. OSWALD, | SIR WILLIAM DAWSON, |
| J. S. BILLINGS, M.D., | J. HUGHLINGS JACKSON, M.D. |

*Subscription price, $5.00 per Annum.*

New York: D. APPLETON & CO., 1, 3, & 5 Bond Street.

# THE BREATH, AND THE DISEASES WHICH GIVE IT A FETID ODOR.

## WITH DIRECTIONS FOR TREATMENT.

### By JOSEPH W. HOWE, M. D.,

Late Clinical Professor of Surgery in the Medical Department of the University of New York, etc.

**Fourth edition, revised and corrected. 12mo, 108 pp. Cloth, $1.00.**

"This little volume well deserves the attention of physicians, to whom we commend it most highly."—*Chicago Medical Journal.*

"To any one suffering from the affection, either in his own person or in that of his intimate acquaintances, we can commend this volume as containing all that is known concerning the subject, set forth in a pleasant style."—*Philadelphia Medical Times.*

# THE NEW YORK MEDICAL JOURNAL VISITING-LIST AND COMPLETE POCKET ACCOUNT-BOOK.

## Prepared by CHARLES H. SHEARS, A. M., M. D.

### Price, $1.25.

This List is based upon an entirely new plan, the result of an effort to do away with the defective method of keeping accounts found in all visiting-lists hitherto published. Each page is arranged for the accounts of three patients, to the number of thirty-one visits each, which may have been made during a current month or may extend over a number of months, according to the frequency of the visits. By this means the necessity for writing a patient's name at each visit, and for searching through several closely-written pages to ascertain how many visits have been made, is obviated. With the simple system here inaugurated, the practitioner can at a glance, and without the trouble of tracing the narrow columns found in the ordinary lists, ascertain the condition of the account of any patient; when, and how many visits have been made; what has been paid, and how much is still due.

It is provided with an Index, and is, without doubt, the most perfect Visiting-List ever offered to the profession, as it possesses all the advantages without the objectionable features found in all others. It can be commenced at any time.

". . . As a convenience and a saver of time by its compactness, this book is worthy of commendation. It seems to us that it would be of great advantage to such as employ it."—*American Lancet.*

". . . Its uniqueness will make it a favorite."—*Practice.*

". . . It forms in itself a complete pocket account-book."—*Northwestern Lancet.*

"This seems the most compendious and convenient account-book for the busy practitioner that could be imagined. . . ."—*Philadelphia Medical Times.*

New York: D. APPLETON & CO., Publishers, 1, 3, & 5 Bond Street.

# MEDICAL

AND

# HYGIENIC WORKS

PUBLISHED BY

*D. APPLETON & CO., 1, 3, & 5 Bond Street, New York.*

---

BARKER (FORDYCE). On Sea-Sickness. A Popular Treatise for Travelers and the General Reader. Small 12mo. Cloth, 75 cents.

BARKER (FORDYCE). On Puerperal Disease. Clinical Lectures delivered at Bellevue Hospital. A Course of Lectures valuable alike to the Student and the Practitioner. Third edition. 8vo. Cloth, $5.00; sheep, $6.00.

BARTHOLOW (ROBERTS). A Treatise on Materia Medica and Therapeutics. **Seventh edition.** Revised, enlarged, and adapted to "The New Pharmacopœia." 8vo. Cloth, $5.00; sheep, $6.00.

BARTHOLOW (ROBERTS). A Treatise on the Practice of Medicine, for the Use of Students and Practitioners. **Sixth edition,** revised and enlarged. 8vo. Cloth, $5.00; sheep, $6.00.

BARTHOLOW (ROBERTS). On the Antagonism between Medicines and between Remedies and Diseases. Being the Cartwright Lectures for the Year 1880. 8vo. Cloth, $1.25.

BASTIAN (H. CHARLTON). Paralysis: Cerebral, Bulbar, and Spinal. Illustrated. Small 8vo. Cloth, $4.50.

BASTIAN (H. CHARLTON). The Brain as an Organ of the Mind. 12mo. Cloth, $2.50.

BELLEVUE AND CHARITY HOSPITAL REPORTS. Edited by W. A. Hammond, M. D. 8vo. Cloth, $4.00.

BENNET (J. H.). On the Treatment of Pulmonary Consumption, by Hygiene, Climate, and Medicine. Thin 8vo. Cloth, $1.50.

BILLINGS (F. S.). The Relation of Animal Diseases to the Public Health, and their Prevention. 8vo. Cloth, $4.00.

BILLROTH (THEODOR). General Surgical Pathology and Therapeutics. A Text-Book for Students and Physicians. Translated from the tenth German edition, by special permission of the author, by Charles E. Hackley, M. D. **Fifth American edition, revised and enlarged.** 8vo. Cloth, $5.00; sheep, $6.00.

BRAMWELL (BYROM). Diseases of the Heart and Thoracic Aorta. Illustrated with 226 Wood-Engravings and 68 Lithograph Plates—showing 91 Figures—in all 317 Illustrations. 8vo. Cloth, $8.00; sheep, $9.00.

BRYANT (JOSEPH D.). A Manual of Operative Surgery. **New edition, revised and enlarged.** 793 Illustrations. 8vo. Cloth, $5.00; sheep, $6.00

BUCK (GURDON). Contributions to Reparative Surgery, showing its Application to the Treatment of Deformities produced by Destructive Disease or Injury; Congenital Defects from Arrest or Excess of Development; and Cicatricial Contractions following Burns. Illustrated by Thirty Cases and fine Engravings. 8vo. Cloth, $3.00.

BURT (STEPHEN SMITH). Exploration of the Chest in Health and Disease. Illustrated. 8vo. Cloth, $1.50.

CAMPBELL (F. R.). The Language of Medicine. A Manual giving the Origin, Etymology, Pronunciation, and Meaning of the Technical Terms found in Medical Literature. 8vo. Cloth, $3.00.

CARPENTER (W. B.). Principles of Mental Physiology, with their Application to the Training and Discipline of the Mind, and the Study of its Morbid Conditions. 12mo. Cloth, $5.00.

CARTER (ALFRED H.). Elements of Practical Medicine. **Third edition,** revised and enlarged. 12mo. Cloth, $3.00.

CASTRO (D'OLIVEIRA). Elements of Therapeutics and Practice according to the Dosimetric System. 8vo. Cloth, $4.00.

COOLEY. Cyclopædia of Practical Receipts, and Collateral Information in the Arts, Manufactures, Professions, and Trades, including Medicine, Pharmacy, and Domestic Economy. Designed as a Comprehensive Supplement to the Pharmacopœia, and General Book of Reference for the Manufacturer, Tradesman, Amateur, and Heads of Families. **Sixth edition,** revised and partly rewritten by Richard V. Tuson. With Illustrations. 2 vols., 8vo. Cloth, $9.00.

CORNING (J. L.). Brain Exhaustion, with some Preliminary Considerations on Cerebral Dynamics. Crown 8vo. Cloth, $2.00.

CORNING (J. L.). Local Anæsthesia in General Medicine and Surgery. Being the Practical Application of the Author's Recent Discoveries. With Illustrations. Small 8vo. Cloth, $1.25.

DOTY (ALVAH H.). A Manual of Instruction in the Principles of Prompt Aid to the Injured. Designed for Military and Civil Use. 96 Illustrations. 12mo. Cloth, $1.25.

ELLIOT (GEORGE T.). Obstetric Clinic: A Practical Contribution to the Study of Obstetrics and the Diseases of Women and Children. 8vo. Cloth, $4.50.

EVANS (GEORGE A ). Hand-Book of Historical and Geographical Phthisiology. With Special Reference to the Distribution of Consumption in the United States. 8vo. Cloth, $2.00.

EVETZKY (ETIENNE). The Physiological and Therapeutical Action of Ergot. Being the Joseph Mather Smith Prize Essay for 1881. 8vo. Limp cloth, $1.00.

FLINT (AUSTIN). Medical Ethics and Etiquette. Commentaries on the National Code of Ethics. 12mo. Cloth, 60 cents

FLINT (AUSTIN). Medicine of the Future. An Address prepared for the Annual Meeting of the British Medical Association in 1886. With Portrait of Dr. Flint. 12mo. Cloth, $1.00.

FLINT (AUSTIN, Jr.). Text-Book of Human Physiology; designed for the Use of Practitioners and Students of Medicine. Illustrated with three hundred and sixteen Woodcuts and Two Plates. **Fourth edition, revised.** Imperial 8vo. Cloth, $6.00; sheep, $7.00.

FLINT (AUSTIN Jr.). The Physiological Effects of Severe and Protracted Muscular Exercise; with Special Reference to its Influence upon the Excretion of Nitrogen. 12mo. Cloth, $1.00.

FLINT (AUSTIN, Jr.). Physiology of Man. Designed to represent the Existing State of Physiological Science as applied to the Functions of the Human Body. Complete in 5 vols., 8vo. Per vol., cloth, $4.50; sheep, $5.50.

  \*\*\* Vols. I and II can be had in cloth and sheep binding; Vol. III in sheep only. Vol. IV is at present out of print.

FLINT (AUSTIN, Jr.). The Source of Muscular Power. Arguments and Conclusions drawn from Observation upon the Human Subject under Conditions of Rest and of Muscular Exercise. 12mo. Cloth, $1.00.

FLINT (AUSTIN, Jr.). Manual of Chemical Examinations of the Urine in Disease; with Brief Directions for the Examination of the most Common Varieties of Urinary Calculi. Revised edition. 12mo. Cloth, $1.00.

FOSTER (FRANK P.). Illustrated Encyclopædic Medical Dictionary, being a Dictionary of the Technical Terms used by Writers on Medicine and the Collateral Sciences in the Latin, English, French, and German Languages. This work will be completed in four volumes. (*Sold by subscription only.*) Volume one now ready. 8vo. Sheep, $10.00; half morocco, $11.00.

FOTHERGILL (J. MILNER). Diseases of Sedentary and Advanced Life. 8vo. Cloth, $2.00.

FOURNIER (ALFRED). Syphilis and Marriage. Translated by P. Albert Morrow, M. D. 8vo. Cloth, $2.00; sheep, $3.00.

FREY (HEINRICH). The Histology and Histochemistry of Man. A Treatise on the Elements of Composition and Structure of the Human Body. Translated from the fourth German edition by Arthur E. J. Barker, M. D., and revised by the author. With 608 Engravings on Wood. 8vo. Cloth, $5.00; sheep, $6.00.

FRIEDLANDER (CARL). The Use of the Microscope in Clinical and Pathological Examinations. Second edition, enlarged and improved, with a Chromo-lithograph Plate. Translated, with the permission of the author, by Henry C. Coe, M. D. 8vo. Cloth, $1.00.

GARMANY (JASPER J.). Operative Surgery on the Cadaver. With Two Colored Diagrams showing the Collateral Circulation after Ligatures of Arteries of Arm, Abdomen, and Lower Extremity. Small 8vo. Cloth, $2.00.

GERSTER (ARPAD G.). The Rules of Aseptic and Antiseptic Surgery. A Practical Treatise for the Use of Students and the General Practitioner. Illustrated with over two hundred fine Engravings. 8vo. Cloth, $5.00; sheep, $6.00.

GROSS (SAMUEL W.). A Practical Treatise on Tumors of the Mammary Gland. Illustrated. 8vo. Cloth, $2.50.

GUTMANN (EDWARD). The Watering-Places and Mineral Springs of Germany, Austria, and Switzerland. Illustrated. 12mo. Cloth, $2.50.

HAMMOND (W. A.). A Treatise on Diseases of the Nervous System. **Eighth edition,** rewritten, enlarged, and improved. 8vo. Cloth, $5.00; sheep, $6.00.

HAMMOND (W. A.). A Treatise on Insanity, in its Medical Relations. 8vo. Cloth, $5.00; sheep, $6.00.

HAMMOND (W. A.). Clinical Lectures on Diseases of the Nervous System. Delivered at Bellevue Hospital Medical College. Edited by T. M. B. Cross, M. D. 8vo. Cloth, $3.50.

4

HARVEY (A.). First Lines of Therapeutics. 12mo. Cloth, $1.50.

HOFFMANN-ULTZMANN. Analysis of the Urine, with Special Reference to Diseases of the Urinary Apparatus. By M. B. Hoffmann, Professor in the University of Gratz; and R. Ultzmann, Tutor in the University of Vienna. Third edition, revised and enlarged. 8vo. Cloth, $2.00.

HOWE (JOSEPH W.). Emergencies, and how to treat them. Fourth edition, revised. 8vo. Cloth, $2.50.

HOWE (JOSEPH W.). The Breath, and the Diseases which give it a Fetid Odor. With Directions for Treatment. Second edition, revised and corrected. 12mo. Cloth, $1.00.

HUEPPE (FERDINAND). The Methods of Bacteriological Investigation. Written at the request of Dr. Robert Koch. Translated by Hermann M. Biggs, M.D. Illustrated. 8vo. Cloth, $2.50.

JACCOUD (S.). The Curability and Treatment of Pulmonary Phthisis. Translated and edited by Montagu Lubbock, M.D. 8vo. Cloth, $4.00.

JONES (H. MACNAUGHTON). Practical Manual of Diseases of Women and Uterine Therapeutics. For Students and Practitioners. 188 Illustrations. 12mo. Cloth, $3.00.

JOURNAL OF CUTANEOUS AND GENITO-URINARY DISEASES. Published Monthly. Edited by Prince A. Morrow, A. M., M D, and John A. Fordyce, M.D. Price, $2.50 per annum, or, if taken in connection with the "New York Medical Journal" ($3.00 per annum), the two publications will be furnished at $7.00 per annum.

KEYES (E. L.). A Practical Treatise on Genito-Urinary Diseases, including Syphilis. Being a new edition of a work with the same title. by Van Buren and Keyes. Almost entirely rewritten. 8vo. With Illustrations. Cloth, $5.00; sheep, $6.00.

KEYES (E. L.). The Tonic Treatment of Syphilis, including Local Treatment of Lesions. 8vo. Cloth, $1.00.

KINGSLEY (N. W.). A Treatise on Oral Deformities as a Branch of Mechanical Surgery. With over 350 Illustrations. 8vo. Cloth, $5.00; sheep, $6.00.

LEGG (J. WICKHAM). On the Bile, Jaundice, and Bilious Diseases. With Illustrations in Chromo-Lithography. 8vo. Cloth, $6.00; sheep, $7.00.

LITTLE (W. J.). Medical and Surgical Aspects of In-Knee (Genu-Valgum): its Relation to Rickets, its Prevention, and its Treatment, with and without Surgical Operation. Illustrated by upward of Fifty Figures and Diagrams. 8vo. Cloth, $2.00.

LORING (EDWARD G.). A Text-Book of Ophthalmoscopy. Part I. The Normal Eye, Determination of Refraction, and Diseases of the Media. With 131 Illustrations, and 4 Chromo-Lithographs. 8vo. Cloth, $5.00.

LUSK (WILLIAM T.). The Science and Art of Midwifery. With 246 Illustrations. Second edition, revised and enlarged. 8vo. Cloth, $5.00; sheep, $6.00.

LUYS (J.). The Brain and its Functions. With Illustrations. 12mo. Cloth $1.50.

MARKOE (T. M.). A Treatise on Diseases of the Bones. With Illustrations. 8vo. Cloth, $4.50.

MAUDSLEY (HENRY). Body and Mind: an Inquiry into their Connection and Mutual Influence, specially in reference to Mental Disorders. An enlarged and revised edition, to which are added Psychological Essays. 12mo. Cloth, $1.50.

MAUDSLEY (HENRY). Physiology of the Mind. Being the first part of a third edition, revised, enlarged, and in great part rewritten, of "The Physiology and Pathology of the Mind." 12mo. Cloth, $2.00.

MAUDSLEY (HENRY). Pathology of the Mind. Third edition. 12mo. Cloth, $2.00.

MAUDSLEY (HENRY). Responsibility in Mental Disease. 12mo. Cloth, $1.50.

MILLS (WESLEY). A Text-Book of Animal Physiology, with Introductory Chapters on General Biology and a full Treatment of Reproduction for Students of Human and Comparative Medicine. 8vo. With 505 Illustrations. Cloth, $5.00; sheep, $6.00.

NEFTEL (WM. B.). Galvano-Therapeutics. The Physiological and Therapeutical Action of the Galvanic Current upon the Acoustic, Optic, Sympathetic, and Pneumogastric Nerves. 12mo. Cloth, $1.50.

NEUMANN (ISIDOR). Hand-Book of Skin Diseases. Translated by Lucius D. Bulkley, M.D. Illustrated by 66 Wood-Engravings. 8vo. Cloth, $4.00; sheep, $5.00.

THE NEW YORK MEDICAL JOURNAL (weekly). Edited by Frank P. Foster, M.D. Terms per annum, $5.00, or, if taken in connection with the Journal of Cutaneous and Genito Urinary Diseases ($4.50 per annum), the two publications will be supplied at $7.00 per annum.

Binding Cases, cloth, 50 cents each.

GENERAL INDEX, from April, 1865, to June, 1876 (23 vols.) 8vo. Cloth, 75 cents.

THE NEW YORK MEDICAL JOURNAL VISITING-LIST AND COMPLETE POCKET ACCOUNT-BOOK. Prepared by Charles H. Shears, M.D. $1.25.

NIEMEYER (FELIX VON). A Text-Book of Practical Medicine, with particular reference to Physiology and Pathological Anatomy. Containing all the author's Additions and Revisions in the eighth and last German edition. Translated by George H. Humphreys, M.D., and Charles E. Hackley, M.D. 2 vols., 8vo. Cloth, $9.00; sheep, $11.00.

NIGHTINGALE'S (FLORENCE) Notes on Nursing. 12mo. Cloth, 75 cents.

PEASLEE (E. R.). A Treatise on Ovarian Tumors: their Pathology, Diagnosis, and Treatment, with reference especially to Ovariotomy. With Illustrations. 8vo. Cloth, $5.00; sheep, $6.00.

PEREIRA'S (Dr.) Elements of Materia Medica and Therapeutics. Abridged and adapted for the Use of Medical and Pharmaceutical Practitioners and Students, and comprising all the Medicines of the British Pharmacopœia, with such others as are frequently ordered in Prescriptions, or required by the Physician. Edited by Robert Bentley and Theophilus Redwood. Royal 8vo. Cloth, $7.00; sheep, $8.00.

PEYER (ALEXANDER). An Atlas of Clinical Microscopy. Translated and edited by Alfred C. Girard, M.D. First American, from the manuscrip of the second German edition, with Additions. Ninety Plates, with 105 Illustrations, Chromo-Lithographs. Square 8vo. Cloth, $6.00.

POMEROY (OREN D.). The Diagnosis and Treatment of Diseases of the Ear. With One Hundred Illustrations. **Second edition,** revised and enlarged. 8vo Cloth, $3.00.

POORE (C. T.). Osteotomy and Osteoclasis, for the Correction of Deformities of the Lower Limbs. 50 Illustrations. 8vo. Cloth, $2.50.

QUAIN (RICHARD). A Dictionary of Medicine, including General Pathology, General Therapeutics, Hygiene, and the Diseases peculiar to Women and Children. By Various Writers. Edited by Richard Quain, M. D., In one large 8vo volume, with complete Index, and 138 Illustrations. (*Sold only by subscription.*) Half morocco, $8.00.

RANNEY (AMBROSE L.). Applied Anatomy of the Nervous System, being a Study of this Portion of the Human Body from a Standpoint of its General Interest and Practical Utility, designed for Use as a Text-Book and as a Work of Reference. **Second edition, revised and enlarged.** Profusely illustrated. 8vo. Cloth, $5.00; sheep, $6.00.

RANNEY (AMBROSE L.). Lectures on Electricity in Medicine, delivered at the Medical Department of the University of Vermont, Burlington. Numerous Illustrations. 12mo. Cloth, $1.00.

RANNEY (AMBROSE L.). Practical Suggestions respecting the Varieties of Electric Currents and the Uses of Electricity in Medicine, with Hints relating to the Selection and Care of Electrical Apparatus. With Illustrations and 14 Plates. 16mo. Cloth, $1.00.

ROBINSON (A. R.). A Manual of Dermatology. Revised and corrected. 8vo. Cloth, $5.00.

ROSCOE-SCHORLEMMER. Treatise on Chemistry.
> Vol. 1. Non-Metallic Elements. 8vo. Cloth, $5.00.
> Vol. 2. Part I. Metals. 8vo. Cloth, $3.00.
> Vol. 2. Part II. Metals. 8vo. Cloth, $3.00.
> Vol. 3. Part I. The Chemistry of the Hydrocarbons and their Derivatives. 8vo. Cloth, $5.00.
> Vol. 3. Part II. The Chemistry of the Hydrocarbons and their Derivatives. 8vo. Cloth, $5.00.
> Vol. 3. Part III. The Chemistry of the Hydrocarbons and their Derivatives. 8vo. Cloth, $3.00.
> Vol. 3. Part IV. The Chemistry of the Hydrocarbons and their Derivatives. 8vo. Cloth, $3.00.
> Vol. 3. Part V. The Chemistry of the Hydrocarbons and their Derivatives. 8vo. Cloth, $3.00.

ROSENTHAL (I.). General Physiology of Muscles and Nerves. With 75 Woodcuts. 12mo. Cloth, $1.50.

SAYRE (LEWIS A.). Practical Manual of the Treatment of Club-Foot. **Fourth edition, enlarged and corrected.** 12mo. Cloth, $1.25.

SAYRE (LEWIS A.). Lectures on Orthopedic Surgery and Diseases of the Joints, delivered at Bellevue Hospital Medical College. **New edition,** illustrated with 324 Engravings on Wood. 8vo. Cloth, $5.00; sheep, $6.00.

7

SCHROEDER (KARL). A Manual of Midwifery, including the Pathology of
Pregnancy and the Puerperal State. Translated into English from the third
German edition, by Charles H. Carter, M. D. With 26 Engravings on Wood.
8vo. Cloth, $3.50; sheep, $4.50.

SCHULTZE (B. S.). The Pathology and Treatment of Displacements of the
Uterus. Translated from the German by Jameson J. Macan, M. A., etc;
and edited by Arthur V. Macan, M. B., etc. With One Hundred and
Twenty Illustrations. 8vo. Cloth, $3.50.

SHOEMAKER (JOHN V.). A Text-Book of Diseases of the Skin. Six
Chromo-Lithographs and numerous Engravings. 8vo. Cloth, $5.00; sheep,
$6.00.

SIMPSON (JAMES Y.). Selected Works: Anæsthesia, Diseases of Women.
3 vols., 8vo. Per volume. Cloth, $3.00; sheep, $4.00.

SIMS (J. MARION). The Story of my Life. Edited by his Son, H. Marion
Sims, M. D. With Portrait. 12mo. Cloth, $1.50.

SKENE (ALEXANDER J. C.). A Text-Book on the Diseases of Women.
Illustrated with two hundred and fifty-four Illustrations, of which one
hundred and sixty-five are original, and nine chromo-lithographs. (Sold by
subscription only.) 8vo. Cloth, $6.00; sheep, $7.00.

SMITH (EDWARD). Foods. 12mo. Cloth, $1.75.

SMITH (EDWARD). Health: A Hand-Book for Households and Schools.
Illustrated. 12mo. Cloth, $1.00.

STEINER (JOHANNES). Compendium of Children's Diseases: a Hand-Book
for Practitioners and Students. Translated from the second German edition.
by Lawson Tait. 8vo. Cloth, $3.50; sheep, $4.50.

STEVENS (GEORGE T.) Functional Nervous Diseases: their Causes and
their Treatment. Memoir for the Concourse of 1881–1883, Académie Royal
de Médecine de Belgique. With a Supplement, on the Anomalies of Refraction and Accommodation of the Eye, and of the Ocular Muscles. Small
8vo. With six Photographic Plates and twelve Illustrations. Cloth, $2.50.

STONE (R. FRENCH). Elements of Modern Medicine, including Principles of
Pathology and of Therapeutics, with many Useful Memoranda and Valuable
Tables of Reference. Accompanied by Pocket Fever Charts. Designed for
the Use of Students and Practitioners of Medicine. In wallet-book form,
with pockets on each cover for Memoranda, Temperature Charts, etc.
Roan, tuck, $2.50.

STRECKER (ADOLPH). Short Text-Book of Organic Chemistry. By Dr.
Johannes Wislicenus. Translated and edited, with Extensive Additions, by
W. H. Hodgkinson and A. J. Greenaway. 8vo. Cloth, $5.00.

STRÜMPELL (ADOLPH). A Text-Book of Medicine, for Students and Practitioners. With 111 Illustrations. 8vo. Cloth, $6.00; sheep, $7 00.

SWANZY (HENRY R.). A Hand-Book of the Diseases of the Eye, and their
Treatment. With 122 Illustrations, and Holmgren's Tests for Color-Blind-
ness. Crown 8vo. Cloth, $3.00.

TRACY (ROGER S.). The Essentials of Anatomy, Physiology, and Hygiene.
12mo. Cloth, $1.25.

TRACY (ROGER S.). Hand-Book of Sanitary Information for Householders.
Containing Facts and Suggestions about Ventilation, Drainage, Care of Con-
tagious Diseases, Disinfection, Food, and Water. With Appendices on Dis-
infectants and Plumbers' Materials. 16mo. Cloth, 50 cents.

8

TRANSACTIONS OF THE NEW YORK STATE MEDICAL ASSOCIA-
TION, VOL. I. Being the Proceedings of the First Annual Meeting of the
New York State Medical Association, held in New York, November 18, 19,
and 20, 1884. Small 8vo. Cloth, $5.00.

TYNDALL (JOHN). Essays on the Floating Matter of the Air, in Relation to
Putrefaction and Infection. 12mo. Cloth. $1.50.

ULTZMANN (ROBERT). Pyuria, or Pus in the Urine, and its Treatment.
Translated by permission, by Dr. Walter B. Platt. 12mo. Cloth, $1.00.

VAN BUREN (W. H.). Lectures upon Diseases of the Rectum, and the Surgery of the Lower Bowel, delivered at Bellevue Hospital Medical College.
**Second edition, revised and enlarged.** 8vo. Cloth, $3.00; sheep, $4.00.

VAN BUREN (W. H.). Lectures on the Principles and Practice of Surgery.
Delivered at Bellevue Hospital Medical College. Edited by Lewis A. Stim-
son, M. D. 8vo. Cloth, $4.00; sheep, $5.00.

VOGEL (A.). A Practical Treatise on the Diseases of Children. Translated
and edited by H. Raphael, M. D. **Third American from the eighth German edition, revised and enlarged.** Illustrated by six Lithographic Plates. 8vo.
Cloth, $4.50; sheep, $5.50.

VON ZEISSL (HERMANN). Outlines of the Pathology and Treatment of
Syphilis and Allied Venereal Diseases. **Second edition,** revised by Maximil-
ian von Zeissl. Authorized edition. Translated, with Notes, by H. Raphael, M. D. 8vo. Cloth, $4.00; sheep, $5.00.

WAGNER (RUDOLF). Hand-Book of Chemical Technology. Translated and
edited from the eighth German edition, with extensive Additions, by William
Crookes. With 336 Illustrations. 8vo. Cloth, $5.00.

WALTON (GEORGE E.). Mineral Springs of the United States and Canadas.
Containing the latest Analyses, with full Description of Localities, Routes,
etc. **Second edition, revised and enlarged.** 12mo. Cloth, $2.00.

WEBBER (S. G.). A Treatise on Nervous Diseases: Their Symptoms and
Treatment. A Text-Book for Students and Practitioners. 8vo. Cloth, $3.00.

WEEKS (CLARA S.). A Text-Book of Nursing. For the Use of Training-
Schools, Families, and Private Students. 12mo. With 13 Illustrations,
Questions for Review and Examination, and Vocabulary of Medical Terms.
12mo. Cloth, $1.75.

WELLS (T. SPENCER). Diseases of the Ovaries. 8vo. Cloth, $4.50.

WORCESTER (A.). Monthly Nursing. **Second edition, revised.** Cloth, $1.25.

WYETH (JOHN A.). A Text-Book on Surgery: General, Operative, and Mechanical. Profusely illustrated. (*Sold by subscription only.*) 8vo. Buck-
ram, uncut edges, $7.00; sheep, $8.00; half morocco, $9.50.

WYLIE (WILLIAM G.). Hospitals: Their History, Organization, and Con-
struction. 8vo. Cloth, $2.50.

www.ingramcontent.com/pod-product-compliance
Lightning Source LLC
Chambersburg PA
CBHW021514210326
41599CB00012B/1251